Sharks of the Shallows

Sharks
of the Shallows
COASTAL SPECIES IN FLORIDA AND THE BAHAMAS

Jeffrey C. Carrier

Photographs by Andy Murch,
Jillian Morris, and Duncan Brake

JOHNS HOPKINS UNIVERSITY PRESS   BALTIMORE

© 2017 Johns Hopkins University Press
All rights reserved. Published 2017
Printed in China on acid-free paper
9 8 7 6 5 4 3 2 1

Johns Hopkins University Press
2715 North Charles Street
Baltimore, Maryland 21218-4363
www.press.jhu.edu

Library of Congress Cataloging-in-Publication
Data

Names: Carrier, Jeffrey C.
Title: Sharks of the shallows : coastal species in
    Florida and the Bahamas / Jeffrey C. Carrier ;
    photographs by Andy Murch, Jillian Morris,
    and Duncan Brake.
Description: Baltimore : Johns Hopkins
    University Press, 2017. | Includes
    bibliographical references and index.
Identifiers: LCCN 2016044752| ISBN
    9781421422947 (hardcover : alk. paper) | ISBN
    9781421422954 (electronic) | ISBN 1421422948
    (hardcover : alk. paper) | ISBN 1421422956
    (electronic)
Subjects: LCSH: Sharks—Florida—
    Identification. | Sharks—Bahamas—
    Identification.
Classification: LCC QL638.9 .C327 2017 | DDC
    597.3/3—dc23 LC record available at https://
    lccn.loc.gov/2016044752

A catalog record for this book is available from
the British Library.

Special discounts are available for bulk purchases
of this book. For more information, please contact
Special Sales at 410-516-6936 or specialsales@
press.jhu.edu.

Johns Hopkins University Press uses
environmentally friendly book materials,
including recycled text paper that is composed
of at least 30 percent post-consumer waste,
whenever possible.

Contents

Preface

So how do you become a marine biologist and study sharks your entire life? Sometimes it turns out to be easier than one would think. I grew up on the northeastern coast of Florida near the fishing village of Mayport, where the St. Johns River empties into the ocean, and it was the perfect setting for an inquisitive youngster with an interest in the ocean. In the 1960s the river and the immediate offshore areas were teeming with fish. The surrounding coastal habitats included miles and miles of salt marsh, rich with nutrients, eventually to be drained by the river's flow and carried into the ocean. The resulting food web was rich and diverse and provided an endless variety of strange creatures to capture and study.

Shrimp boats were common and the Mayport fisheries thrived. Wherever high-nutrient flow, invertebrates, and small fish abound, larger fish are certainly nearby. These top-level predators included king mackerel, bonitos, and, of course, many species of sharks. Bait could be obtained from the bycatch of shrimpers' nets, enough for a full day of fishing. A washtub could be filled with endless varieties of fish, crabs, and all sorts of bizarre eels for a dollar. After getting bait, we would anchor off the jetties at the mouth of the St. Johns and throw out the chum, hoping to bring home a kingfish to appease our parents, who had only a meager understanding of the risks of offshore fishing

from a 14-foot boat with an 18-horsepower engine. More often than not, however, our catch was dominated by sharks. Many years later I learned that the number and diversity of sharks in this area explained why the US Navy chose Mayport to develop and test shark repellents during the Second World War.

In the early sixties, partly because of the popular music of the time (think "Beach Boys" and "Jan and Dean"), having a surfboard and riding what passed for waves in northern Florida was every young person's passion. Riding the waves invariably brought us into the realm inhabited by sharks, mostly small and almost always blacktips or spinners. These encounters never led to problems, and we seemed to coexist easily, enjoying the sun, surf, and countless fish in the nearshore waves.

Natural curiosity got the best of us, so we began to fish these waters in between surfing expeditions. Fishing in the surf was not what we were used to and required different equipment and skills. Many fishers, rather than fishing the surf among the crowds of beachgoers and surfers on the beach, chose the Jacksonville Beach pier, which provided a bit of a haven. It extended off the beach just enough to escape the waves and made it easier to land whiting, bluefish, and an occasional kingfish. The crowd on the pier almost always included diehard shark fishers, so it became a place for us to see these critters close up and learn how they could be caught.

Regrettably, the most common outcome for the sharks was to be bludgeoned to death by the fishers and then discarded, a wasteful practice that reflected the low regard for these animals. Knowing that the sharks would be thrown away, several of my intrepid buddies and I would ask for the carcasses so that we could look inside the dead fish; this activity was regarded by the pier fishers as ghoulish behavior. Since we would then be the ones to have to clean up the mess, they were content to shift those duties to us. But for curious teenagers, the experiences were mesmerizing. A 12-foot hammerhead once yielded 21 pups, and we were hooked forever.

We soon began to hone our own fishing skills, pooling our lawn-mowing money to buy one heavy rod and reel to be shared, and branching out along the coast from Amelia Island and Little Talbot Island to the north, to Matanzas Inlet near St. Augustine and the Shark Pit in Melbourne, where we could combine our interests in shark fishing and surfing on what were regarded as the best surfing beaches along Florida's east coast. Our expeditions were all occasioned by the Paul Maines column *Fins, Fur, and Feathers* in the *Florida Times Union,* which reported on the Florida Shark Club's outings. We followed Maines, through his column, as though he was the pied piper of all creatures with fins, and we sang his written tunes as we pursued redfish, tarpon, king mackerel, and of course, sharks.

About the same time, we became interested in a new sport. Fascinated by the television adventures of the character Mike Nelson, a former navy frogman played by Lloyd Bridges on the television series *Sea Hunt,* we decided we should add scuba diving to our growing repertoire of ways to dodge homework and jobs. There were wonderful offshore artificial reefs off Mayport, and on one of these reefs, after a particularly bad day of fishing, I made my first scuba dive in 1962. The very

first fish I encountered on an 80-foot dive was a small shark. I knew very little of these fish and believed at the time that the 3-foot animal was preparing to attack and devour my best friend and me. But the shark swam away almost as fast as we did. When we returned home and finished cleaning the boat, I immediately went to our local library and checked out every book about sharks I could find, so I could understand how I survived this harrowing experience. It turned out that there was almost nothing known about these mysterious animals.

My experience was probably no different from that of many others who venture into Florida waters. For visitors to Florida or the Bahamas, or for the residents, the chances of encountering a shark are very high and the sight of a shark in the water is very common. For recreational boaters, it is generally a chance encounter, one that might occur in very shallow water near the shoreline. For divers, who immerse themselves in the watery world that surrounds the Florida coastline and the out-islands of the Bahamas, the likelihood is much greater. And for anglers who place bait in the water for other fish, the sight of a shark, attracted to their bait or to a fish on their line, is even more likely.

Soon the carefree days of high school ended, and for most of my fishing buddies, college or workdays began to interfere with sun and surf. For me, after a year of college, the confines of the classroom were too much, so I decided to satisfy my military obligation (in my youth, military service was not optional) and joined the Coast Guard. My duty assignment took me to Tampa Bay, where I saw schools of manta rays near the base of the Sunshine Skyway bridge and small sharks in the shoal waters around Egmont Key, all from the deck of a 40-foot patrol boat. I finished active duty aboard a buoy tender out of Mayport and was thrilled to see large sharks near the river mouths as we serviced and replaced the sea buoys along the Florida and Georgia coasts, waters that were familiar to me.

When I returned from active duty, I knew that all I wanted to do was fish for sharks. I merely needed to find a way to get paid to fish and have someone else pay for the bait. It was clear that I would need formal marine bio training, and my main requirement was that the studies had to include sharks. The University of Miami became my destination to complete undergraduate and graduate studies in marine biology. The first class I signed up for was a fish biology class taught by Dr. David Evans. His specialty was salt and water balance in marine fish. For the required term paper, I chose to write on—what else?—salt and water balance in sharks. There was essentially nothing known at the time about how sharks managed their internal salt content in seawater, so when the course was completed, I proposed an independent study to investigate how sharks managed to regulate this vital physiological function. Dr. Evans had significant grant funds, so criterion number one, someone else buys the bait, was satisfied. I then seriously began to fish for sharks, albeit interrupted considerably by the requirement to conduct the basic research for my graduate studies, and a lifelong commitment to the formal study of sharks began.

While many people who spend a great deal of time on the water are generally familiar with sharks, very few understand the complex biology of these animals, and very few can successfully identify more than a few species. At no time in our history has the understanding of these animals and the ability to identify them been more important. Many species have been fished to near extinction and are now protected. Other species are not under any special stress, and it is important to be able to distinguish which ones are protected. It is equally important to understand how it is that these animals are unable to tolerate heavy fishing pressures and why many of their populations might be in trouble.

It was with these concerns in mind that I wrote this book: to help marine travelers identify the sharks they encounter and understand the biology of this important and exciting group of fish. It is written for people who probably have some basic understanding of sharks. But it is not an attempt to provide all the information ever written about sharks. Furthermore, the animals that are described are those that are most likely to be encountered in the nearshore, shallow waters that surround Florida and the Bahamas. No book could hope to cover all of the more than 500 known species in depth. Nor does this book include every single species that has ever been described in Florida waters. For example, many deep-water species have been found offshore but are not likely to be captured or encountered by recreational boaters, so they have been intentionally omitted. While the book will aid in identifying sharks that might be encountered, it is not intended primarily as a key to identification. I see it mostly as a key to understanding these species and secondarily to aid with identification. It is not written with the scientist in mind and it is not a textbook; there are already many such texts. This book is for people who spend time on, in, or under the water, those who would like to learn more about sharks.

The book is divided into three parts. The first part presents a substantial overview of shark biology for the layperson, along with basic clues to shark identification. Part 2 describes selected species and provides keys to identifying each one as well as a brief indication of how common that species is. Please read the "frequency" comments with the realization that some sharks considered to be common in one area may be rare in others and that the frequency of some sharks may be seasonally dependent and influenced by migratory movements. Part 3 offers information about some of the relatives of sharks that we often find with, or instead of, sharks.

Acknowledgments

A book of this scope is seldom the work of one individual. I have been so fortunate as to work with many of the most important shark scientists, so I can't possibly list everyone whom I would like to acknowledge. Perry Gilbert and Eugenie Clark, both legends in the field of shark research and both now deceased, were certainly my earliest influences. Wes Pratt is a long-time colleague with whom I have studied shark reproduction and mating behaviors for more than two decades; he and I have often ventured into discussions very unrelated to sharks, and he has been inspirational in ways that only a true friendship can foster.

Mike Heithaus, a dynamic young scholar with whom I have coedited many books, written many papers and grant proposals, and fished side-by-side for sharks on countless expeditions, is one of the finest writers and teachers that I have been privileged to work with. Carl Luer is the consummate professional and probably the hardest-working, most professional biologist I have ever been associated with. He may also be the most respected colleague I know, and I prize my friendship with Carl.

Jack Musick, Samuel "Doc" Gruber, Bob Hueter, Colin Simpfendorfer, Michelle Heuple, Sonja Fordham, Janine Caira, Greg Cailliet, José Castro, George Burgess, and Gene Helfman have each, through the years, studied problems of common interest along with me and have all been influential in one way or another. Frank Murru, Gary Violetta, Ray Davis, and Forrest Young are professionals from the world of zoos and aquariums who know more than is imaginable about safe collecting, animal husbandry, and ways to keep these animals alive and healthy.

My sister Patti has always been there and has always been excited about and interested in my work. A shark fanatic in her own right, she is probably more skilled at finding shark teeth on the beaches than anyone else in northern Florida. Her gallons and gallons of teeth are testimony to her unique skills and passion for the quest.

My research has always been based in the Florida Keys. The earliest years were spent learning the waters of the Keys and the particular hangouts of nurse sharks. No one was better acquainted with the shallow inshore waters or knew more of the habits of these sharks than Captain Billy Schwicker. We spent years on the flats together, and he invested innumerable hours schooling me in the ways of these wonderful animals, interspersed with tall tales of the tropics, an inspiration in their own right—though the truth was sometimes as elusive as the nurse sharks.

The facilities of the Newfound Harbor Marine Institute at Seacamp have provided a base of operations and willing personnel and logistical support for my entire career. Irene Hooper, Grace Upshaw, and the hundreds of students and staff who have weighed, measured, and tagged sharks as part of my ongoing research have made much of my work possible. Buffy Redsecker and Alan Chung have very graciously offered the ideal setting where much of the creation of this book occurred.

My students who have chosen paths in shark biology include Nick Whitney and Derek Burkholder, both of whom, as scientists with connections and deep ties to Albion College, hold much promise to continue the study of sharks. Finding the literature and the scientific papers that are required reading for a scientist is often challenging. Librarian Mike Van Houten is a genius when it comes to tracking down the important work done in the field of shark biology. I am deeply indebted to him for his good humor and exceptional professionalism.

I am especially grateful to my photographers, Jill Morris, Andy Murch, and Duncan Brake. The quality and professionalism of their work is obvious on nearly every page of this book. They are knowledgeable, professional, and absolutely intrepid in their quest for just the right image. The photos on the following pages are testaments to their successes, and Duncan's illustrations are true works of art. It has been a joy to work with these three and learn from them as we brought this volume to completion.

My lovely wife, Carol, has tolerated my field seasons and months of absences as only a truly supportive spouse could. She has worked in the field with me, has handled small-boat operations, has lugged gear everywhere, has helped with surface photography and videos, and has been indispensable with her support and encouragement. She has seen me through many books and papers, and I constantly marvel at her tolerance and her spirit. It is to her that I dedicate this book and the many adventures to follow.

Sharks of the Shallows

PART ONE

Introduction to Sharks and Their Relatives

The sight of a shark in the water evokes a wide range of emotions, ranging from simply being startled, to wild excitement, paralyzing fear, and sheer fascination. How you react probably depends on your past experience (if any) with sharks, your age, and your natural curiosity.

What does it have to do with your age? For people older than 50 or 60, the thought of sharks may bring to mind the popular book and movie titled *Jaws,* which, back in the mid-1970s, featured a large, revenge-minded shark with a taste for human flesh. It was a creation of fantasy, but it evoked primal fears of being attacked and eaten by a mindless killing machine. Forgotten is the fact that *Jaws* was a work of fiction by a very skillful writer, Peter Benchley. For many years following the appearance of *Jaws,* Benchley worked tirelessly to convince readers and moviegoers that his epic story was, in fact, a fictional creation. He became passionately involved with conservation efforts, as though he was doing penance for creating the alarming negative fervor that was an unexpected consequence of his book and movie.

But everyone is also curious, either immediately or after the initial shock of encountering a shark wears off: Why is this animal suddenly here? Is it dangerous? Do I have anything to fear? Should I try to scare it away? Should I just leave? Should it be killed? Although younger people, fortunately, do not

The sight of a fin in the water (*top*) is universally understood. It is often the first sign that announces the presence of a shark. It may inspire panic or exhilaration or some other combination of emotions. But it will always result in curiosity, especially when the entire shark comes into view (*bottom*).

Unfortunately for the reputation of sharks, the image of a killing machine was wrongly inspired by movies like *Jaws* and created a fear of sharks.

share their parents' and grandparents' fear of sharks, there is a general lack of knowledge about this group of fish. However, for those of us who have been on, in, around, or under the water and have some level of understanding of and experience with sharks, every encounter is exhilarating. Coexistence is becoming the model for our frequent interactions.

More and more educational programming—based on sound science and exploration rather than sensationalistic portrayals of killer sharks the size of submarines or sharks raining down from the effects of tornados—is providing a sound foundation from which sharks can be seen as important inhabitants of the underwater world. They have clearly delineated roles in their environments. Ecosystems become weaker and are challenged when sharks are removed. This part of the book introduces sharks and some of their relatives with discussions of traits common to all or most species. The text that follows

is devoted to discussing the most common shark species in Florida and Bahamian waters.

Ancestors and the Family Tree

The fossil history of sharks dates back nearly 400 million years to a time known as the Devonian period or, more commonly, the Age of Fishes. The primitive jawless fish began to decrease in abundance and were replaced by fish with jaws, including many primitive species of armored fish and the earliest sharks.

Sharks and their cousins, the skates and rays, lack true bone in their skeletal structure. Instead, their skeletons are made of cartilage. Cartilage does not preserve well over millions of years, so the fossil record is not as complete as it is for bony fish and other vertebrates. Most attempts to reconstruct ancient sharks have used teeth and tooth fragments, mineralized skeletal elements that can withstand deterioration over time.

From these fragments have emerged models of the extinct and well known Megalodon or megatooth shark (*Carcharocles megalodon*), thought to have reached a length of 52 feet (15.8 m), with teeth that may have been almost 7 inches long (18 cm).

Not all fossil reconstructions are sharklike in appearance. In fact, some early fossils do not even look like fish. The jaws and tooth arrays in some recovered fossils are extremely bizarre. One species, *Helicoprion,* had a jaw that looks more like a poorly designed circular saw with giant teeth.

Sharks are perfectly good vertebrates and perfectly good fish. But their lack of bone places them in a different group from most other fish. They belong to the class Chondrichthyes (cartilaginous fish) rather than the class Osteichthyes, which includes fish that have true bone. Many biologists also refer to the sharklike fish as elasmobranchs. And most classification schemes place the sharklike fish into a group known as the subdivision Selachii and place the skates and rays, sharks' close relatives, into the subdivision Batoidea (often simply referred to as batoids). It is within this latter group that we also find the endangered sawfish, unique and strange fish that are found in Florida and the Bahamas.

Although most people probably have a mental image of sharks, there are many species that do not seem sharklike. The goblin shark or the cookie-cutter shark, with their bizarre and alien features, would not remind anyone of a shark. For a biologist, however, it is not good enough to say that a shark looks like a shark or does not. As everyone knows, biologists are driven by their need to classify things.

Classification

Since the use of common names can sometimes complicate the identification of species, most biologists, fisheries managers, and serious aquarists rely on the scientific names to identify different species. For example, the

Ancient, extinct precursors to modern-day sharks left behind teeth and tooth fragments that hint at the large sizes that these sharks reached. Teeth thought to be from the legendary *Megalodon* may be too large to fit in an adult hand (*top*). Some have been found as large as 7 inches or more in length. Besides individual teeth, bizarre tooth spirals, known as whorls, have been found from an ancient shark known as *Helicoprion* (*bottom*), which is thought to have been extinct for nearly 250 million years. Shark design has clearly undergone substantial changes during sharks' evolutionary history.

term "sand shark" is incorrectly used to describe almost any small shark seen in shallow water. "Sand shark" more correctly refers to the sandtiger shark. These animals are not generally the small coastal species found in the shallow nearshore waters where most visitors claim to have encountered a sand shark. Further, in different parts of the world, the cub shark, the ground shark, the bull shark, the Zambezi River shark, and the Lake Nicaragua shark might be seen. But all these different names apply to only one species, *Carcharhinus leucas,* most commonly known as the bull shark in Florida and the Bahamas.

While the use of Latin names can be cumbersome for nonscientists, it is the only sure way to be certain of a species identification. Every known living species can be described with just two words. Throughout the book I use common names whenever possible. The first time a species is mentioned, I will include the scientific name to avoid confusion and to provide a tool for readers who wish to study a particular species in more detail than is provided in this book.

The bull shark, as it is known in Florida and the Bahamas, has so many common names that understanding what species is being discussed can be an issue. Using scientific nomenclature is the only way to be certain what animal a discussion refers to. *Carcharhinus leucas* is known to all who study this species and avoids the confusion that accompanies the use of local or common names.

The recognizable streamlined, or fusiform, shape of the shark resembles a torpedo and enhances the ability of the shark to swim through the water at high speeds because of the minimal drag that this shape provides.

How to Build a Shark

Most sharks are torpedo-shaped, a design biologists call fusiform. Such a shape enables sharks to fly through the water and overcome the resistance encountered in the aquatic world. Submarines are designed with a similar architecture to enhance their passage through water. For fish, the fusiform design helps them swim with a relatively small energy expenditure. If popular accounts are to be believed, some sharks, notably the shortfin mako shark (*Isurus oxyrhincus*), are among the fastest swimming fish, with burst speeds up to 43 to 46 miles (60 km) per hour. Only several species of billfish and tuna are faster. Sharks' continuous-swimming speeds are considerably less, seldom topping 5 to 10 miles per hour.

Shark fins, another easily recognizable feature, often break the surface of the water to announce sharks' presence to observers. In addition to the dorsal fins on the top of the body and a single anal fin on the ventral surface (underside), sharks possess two sets of paired fins. The fins closest to the head on the sides of the animals are the pectoral fins, and the paired fins nearer the tail are the pelvic fins. These fins may be highly modified in some species. The pectoral fins of skates and rays, the batoids, are broad and extend toward the head and the sides, giving these animals their rounded appearance. The pelvic fins are modified in males as reproductive structures.

Swimming movements by sharks are dependent on the ability of the fins to provide lift through the water as they swim, much like the wings of an aircraft or the bow planes of a submarine. This ability to change their vertical position in the water column is their sole means to rise or sink in the water column, since they have no swim bladder. The dorsal and anal fins are more suited to keeping the shark stable and level in the water. Knowing the relative sizes of the first and second dorsal fin is often useful in identification of sharks. For example, the first and second dorsal fins of a lemon shark (*Negaprion brevirostris*) are almost equal in size, but the first dorsal fin of a Caribbean reef shark (*Carcharhinus perezi*) is much larger than the second dorsal fin.

The tail fin is also referred to as the caudal fin. Its main function is to provide the thrust necessary to move the animal through the water. The tail usually has two lobes, an upper lobe that is generally longer and a lower lobe. This type of tail structure is termed a heterocercal tail, compared to a more crescent-shaped tail, which has upper and lower lobes almost identical in length. This latter shape is called a homocercal tail and is typical of some of the faster-swimming sharks. While propulsion is the principal role of the tail, the thresher sharks (*Alopias* sp.) are able to stun fish with rapid movements of the tail and then eat the stunned fish.

The shape of the tail is also an important key to identifying different sharks. In addition, in some sharks, where the body tapers just before the tail is formed, the region termed the caudal peduncle, there may be ridges that extend out from the peduncle. This is very noticeable in tiger sharks (*Galeocerdo cuvier*) and is yet another feature that helps to distinguish between species.

No Bones about It

What makes a shark a shark? What are the characteristics that make sharks different from the usual kind of fish that we see in the shallow waters and on the coral reefs of Florida and the Bahamas? And, of course, what makes sharks so well suited to be a top-level predator in the marine environment?

The skeletal structure of sharks consists

The pectoral fins of sharks provide the lift required to change their vertical position in the water. Because sharks lack a swim bladder, they must actively swim to go up or down in the water column. The pectoral fins provide lift in the same fashion as the wings of an airplane.

The pectoral fins of the rays are very large and are modified to appear as a part of the body. They play a more active part in propelling the rays than sharks' pectoral fins do for sharks. They actually "ripple" with an undulating movement that helps the ray to swim, since it lacks a tail to propel it through the water.

The relative size of the fins and where they are placed with respect to each other is one way to aid in shark identification. The lemon shark (*Negaprion brevirostris*) has two dorsal fins that are approximately the same size, and this distinguishes it from many of the other sharks of Florida and the Bahamas.

The cartilaginous nature of the skeletons of sharks, skates, and rays is a characteristic of this group of animals. It is clearly visible in these prepared specimens from Dr. Adam Summers of the University of Washington's Friday Harbor Labs: the stained skeleton of a butterfly ray (*Gymnura crebripunctata*) (*top*) and of a bonnethead shark (*Sphyrna tiburo*) (*bottom*). Dr. Summers's work, offering fine examples of biological art, has been displayed in art shows as well as in the scientific literature. *Photographs copyright Adam Summers. Used with permission. All rights reserved.*

of cartilage, not bone. Since they are aquatic creatures, the water provides sufficient buoyancy; no real strength is required of a skeletal system to support the body mass. Cartilage is fully capable of bearing the animal's weight and providing attachment points for muscles and tissues. Cartilage is also thought to have greater flexibility than bone. When the tail is flexed in one direction during swimming movements, this flexibility may help it to snap back and avoid the expenditure of energy to contract muscle in completing the

movement. This may provide an economy of sorts and reduce the amount of energy needed to swim constantly.

How to Breathe under Water

Fish rely on the oxygen dissolved in water to support their metabolic functions, just as terrestrial organisms need oxygen from the air. To keep the oxygen level high so that it may be obtained from the water by the gills, fish must either swim constantly or pull a current of water in through their mouths (the more technical term for mouth is buccal cavity) and across the gills, which are specially adapted organs that extract oxygen from the water. Very few shark species have the musculature to power this pumping mechanism, known as a buccal pump. Nurse sharks, lemon sharks, and some of the smaller cat sharks have these adaptations and are able to rest motionless on the sea floor. Most other species must swim constantly to keep the water near the gills fresh and high in oxygen content, to avoid oxygen depletion and suffocation. This type of ventilation is termed ram ventilation. Sharks that are caught and removed from the water will soon die unless water is passed across their gills in some fashion. Most biological tagging efforts use a seawater hose that is inserted into the mouth of a captured and restrained shark so that the shark can breathe during the measuring and tagging.

Sharks possess gill slits, openings on both sides of the head just behind the mouth. Water that is taken in through the mouth passes across the gills, oxygen is captured by the gills, and the water then passes out of the shark through the gill slits. Most species have five slits, but several have six or seven; they are cleverly called six-gill and seven-gill sharks.

Skates and rays also have gill slits, but they are located on the bottom of the animal. Those fish also possess an additional opening on the top of the head called a spiracle. Water may leave the fish through the spiracle when

Gill slits along the sides of a shark's body allow water to enter through the mouth, pass along the gills, and leave the animal through the gill slits.

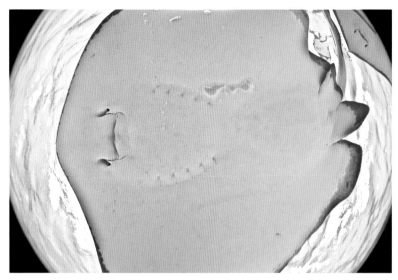

The flattened underside of rays, photographed here with a fish-eye lens, allows them to rest flat on the sea floor. They may flap their pectoral fins, disturbing the sand bottom, and become covered with sand. This camouflage hides them well from potential prey but requires other adaptations to breathe while they are partially buried in the sand.

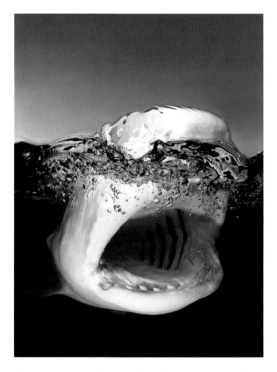

As water enters the mouth and passes across the gills, shown from inside the mouth of this tiger shark (*Galeocerdo cuvier*), oxygen is extracted from the water and waste products leave the body.

The spiracle behind the eyes of this southern ray (*Dasyatis americana*) allows it to breathe. Even though the gill slits are buried under the sand, the spiracle permits water to be drawn across the gills so that adequate oxygen can be provided to support respiration.

the skate or ray is resting on the bottom of the sea. If the animal is partially buried in the sand, the gill slits may be resting on the sand, so that the water cannot move out of the animal. The spiracle solves that problem.

Sink or Swim?

Sharks lack a swim bladder, the internal structure found in bony fish that can be filled and emptied of oxygen from the blood to provide buoyancy and to permit those fish to change their depth. Since sharks are thus negatively buoyant, they sink when they stop swimming. They must therefore utilize other adaptations to achieve buoyancy control. Since most species must swim to breathe

The pectoral fins of this oceanic whitetip shark (*Carcharhinus longimanus*) very closely resemble the wings of an airplane. They function much the same way to provide lift though the water column, compensating for the absence of a swim bladder.

(particularly open-water [pelagic] species), they are able to change their vertical position in the water only through controlled swimming movements and the use of their fins, which serve the same function in the water as airplanes' wings do in the air. In addition, many species produce oils, which are stored in the liver; the oils provide them with additional lift and reduce their tendency to sink. These characteristics of sharks also prevent them from hovering in the water, unless there is a current present, and keep them from turning around in tight circles like fish with swim bladders, which can hover and turn from a stationary position.

Wrapping It All Up: Skin and Scales

Like most fish, sharks have skin that is protected by scales. However, their scales, called placoid scales, do not resemble the circular scales found in most fish (cycloid or ctenoid scales). Instead, they look like microscopic teeth. These "skin teeth" or, more technically, dermal denticles, protect the skin but may also reduce drag when a shark is swimming and thereby provide an additional level of hydrodynamic efficiency. Not surprisingly, bottom-dwelling sharks that spend time under corals or rocky ledges have thickened scales that are designed less for efficient swimming and more to protect against the sharp edges of corals. The toothlike structure of the denticles is not coincidental, for the same tissue that gives rise to the teeth in embryonic sharks also leads to development of the scales. In microscopic cross-section, the denticles even have an internal appearance resembling a tooth with a central cavity.

The skin of female sharks is measurably thicker than that of males. Their courtship and mating behaviors involve bites from male sharks to permit the mating act to occur. Fe-

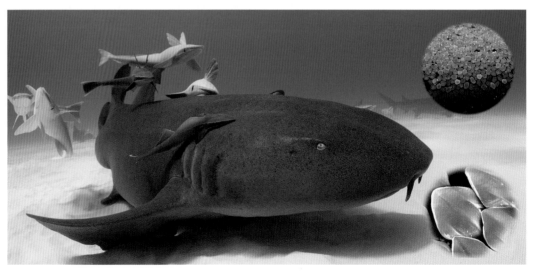

Magnification of shark skin shows the detail of the dermal denticles. The black and white inset shows an extreme magnification that better illustrates the toothlike structure of the scales, which are referred to as skin teeth.

males are frequently discovered with scars from past mating events that could be deadly without the protection afforded by the thickened skin. Even with their thick skin, however, serious injuries can result from attacks by other fish.

Shark skin is so rough that commercial industries have developed techniques to cure the skin with the denticles intact to use it as a fine grade of finish sandpaper called shagreen. If the denticles are removed from the cured skin, a fine grade of leather is produced. Both products illustrate the strength and toughness of shark skin. Brushing against a shark underwater or when a shark is being removed from a fishing line can lead to a nasty scrape because of the rough skin texture.

An Eating Machine?

When people are asked what anatomical feature first comes to mind when they think of sharks, they usually speak of the jaws and teeth. This is not particularly surprising; most people regard sharks as swimming teeth, top-level predators that are "killers of the sea." While sharks certainly are capable hunters that must somehow find and kill prey items to feed on, many species are much more benign hunters. The largest sharks, the whale shark (*Rhincodon typus*), the basking shark (*Cetorhinus maximus*), and the Greenland shark (*Somniosus microcephalus*), are all plankton feeders, relying on the smallest ocean organisms for their nutrition. Many other species are bottom-feeders and opportunistic feeders: small mollusks, crustaceans, and dead material are their food sources.

The chemical composition of teeth is different from that of the cartilaginous skeleton. Some calcification occurs in the teeth and enamel and even in the basal structure of the denticles, and these are all somewhat more resistant to the effects of aging. Thus the fossil record for sharks depends a great deal on what teeth or tooth fragments can be recovered. Occasionally an entire jaw might be found essentially intact. From such remains,

The heavy denticles of a nurse shark (*Ginglymostoma cirratum*) protect it well from the sharp edges of coral reefs, one of its preferred habitats. Even when one uses a scalpel blade to attach a tag, it is very difficult to cut through the skin.

Even with a thick skin and the protection offered by their armored skin teeth, sharks can still be severely wounded by encounters with predators. Whether the scars on this white shark (*Carcharodon carcharias*) originated from combat with another shark or were the result of mating attempts is unclear. The thickness of the skin offers protection against the injuries that occur even to these high-level predators.

forensic paleobiologists have the capability to reconstruct entire animals. It is in this manner that models of the earliest sharks have been reconstructed, and this is how the formidable model of Megalodon, or the Megatooth shark (*Carcharocles megalodon*), was developed. Megalodon, long extinct, was previously thought to be an ancestor of

Whale sharks (*Rhincodon typus*) and the other largest shark species are filter feeders, traveling through swarms of plankton with their mouths open to strain these microscopic organisms from the water. Small fish and crabs, as well as other small food items, are often part of the food they ingest.

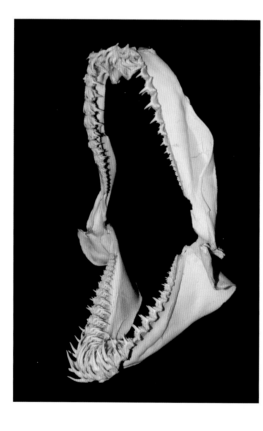

The shark jaw is an engineering marvel, with bite forces that can easily drive a shark's sharp teeth through prey items. Many rows of teeth provide ready replacement for teeth that are lost in violent feeding episodes.

Sharks use a wide variety of feeding strategies. Some are fast-attacking animals that can outswim their prey and consume them whole. Others are well camouflaged, lie on the bottom, and ambush prey that may stray too close. Still others are capable of sucking prey items such as octopuses or other invertebrates out of their hiding place. Nurse sharks can develop such a suction force that they can actually suck a mollusk such as a conch right out of its shell.

For those species that are faster and more accomplished hunters, the jaws and teeth are engineering marvels. Unlike many animals that have one or two sets of teeth throughout their lifetime, sharks continue to produce teeth for as long as they live. The teeth are set in rows within the jaw structure. When a tooth is lost in the sometimes violent feeding behaviors, another will roll forward to replace it in a matter of days, ranging from 2 days to 70 or 80 days, depending on the species.

Some areas of the world are well known for the huge numbers and wide array of teeth that can be found on their beaches. Collecting these beach treasures requires a sharp eye and developing a search image that can detect a tooth on a beach littered with shell debris and other beach litter. Recovered teeth often find their way into jewelry and other trinkets.

Tooth design is variable among the species as well. Some are more pointed, presumably to better hold on to prey items once they have been captured. Others are more capable of cutting or tearing flesh or food that is too large to be swallowed whole. Bottom-dwellers' teeth are wider and flatter and are better designed to crush food such as mollusks in their shell, or crabs, lobsters, and other hard-bodied crustaceans.

The jaw design of many sharks is equally fascinating. Some species can extend their upper jaw, in an action called protrusion. There is not universal agreement on what function this feeding mechanism may have.

great white sharks. But recent fossil evidence seems to suggest a different lineage. It was among the largest sharks ever to roam the seas and may have reached lengths in excess of 50 feet.

Some biologists believe that it aids in impaling prey items more efficiently, so that teeth of the lower jaw may function more effectively to cut and tear the flesh of a prey item. This feature may also improve grasping ability or the ability to feed on food items taken from the bottom. There is total agreement, however, that these animals are effective in their ability to capture prey, regardless of the particular method of feeding.

The food items that have been discovered in the stomachs of sharks are as diverse as the sharks themselves. Fish of every species; squid; octopuses; a huge variety of crustaceans such as crabs, lobsters, and shrimps; and mollusks such as clams and conchs are common table fare for sharks and their skate and ray cousins. While some shark species have preferred food items, many are more opportunistic, feeding on whatever might be available at mealtime.

Digesting License Plates and Baseballs

The digestive system of sharks is perfectly adapted for their role as a top predator. It is fully capable of digesting the various pieces and parts obtained by consuming fish, crustaceans, mollusks, and all manner of aquatic animals. Remains of birds, cows, iguanas, and the occasional license plate have actually been removed from the stomachs of some sharks. Fish, however, make up more than 75% of the food items sharks consume, despite the many oddities that have been reported. The larger sharks, however, rely on plankton, the smallest creatures in the sea, for their diets.

Studies of the digestive processes of sharks have been difficult. It is very nearly impossible to watch a free-swimming animal throughout its daily feeding activities and then to capture its feces to estimate how efficient the digestive system is at extracting nutrition from the animal's meals. Even with animals in captivity, such studies are tricky. The one species in the wild that has been

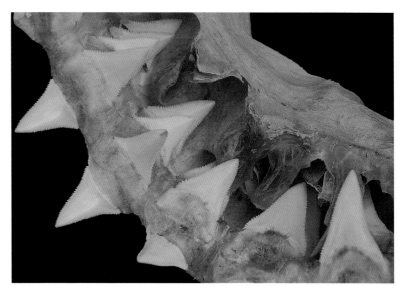

The rows of teeth are readily visible in this jaw from a white shark. Teeth in the back rows can roll forward to replace a lost tooth within days; this adaptation ensures that these predators are always well equipped to feed.

Lost teeth are considered treasures for beachcombers. Collectors along the northeastern coast of Florida frequently fill jars with teeth collected along the surf. Why this particular area is such a treasure trove for teeth is unknown, but it highlights the large numbers of sharks that are found in these waters. *Photograph courtesy Anthony Hodge. Used with permission.*

studied is the lemon shark, whose digestive efficiency was calculated at nearly 80%.

This high rate of efficiency may be due in part to the unique structure of the shark intestine. The inside of the intestine resembles a spiral staircase and is commonly referred to

Blue crabs (*Callinectes sapidus*) may be "beautiful swimmers" to humans, but to sharks they are one of many favored crustacean delights.

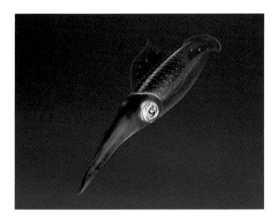

Squid are a primary item in the diets of many of the more pelagic shark species. Many studies of shark stomachs have shown vast quantities of squid, cuttlefish, octopuses, and other cephalopods that make up a large portion of their diets.

The Florida or Caribbean spiny lobster (*Panulirus argus*) is as highly prized by sharks as a food dish as it is to humans. Many divers have had arguments with sharks, particularly nurse sharks, about who actually owns a dive bag full of lobsters. Not all of these arguments end in favor of the diver.

Mysid shrimp, shrimplike crustaceans seldom more than 1 inch long, are a major portion of the diets of whale sharks and many other plankton feeders. Migrations and long-distance movements by these large fish often coincide with huge plankton blooms; large schools of whale sharks congregate to feed at such times.

as a spiral valve. The spiral shape means that the food moves in a spiral, or circular, motion as it passes through the digestive tract. The effect is to bring food into contact with a greater surface area, making both digestion and nutrient absorption more effective. It also means that a shorter total intestinal length can provide an adequate amount of surface area for the digestive processes. The efficiency of this system may be further enhanced by the amount of time a meal spends inside the shark. Some estimates have shown that certain shark species may take upwards of 20 days to digest and eliminate a meal. If

this is true, it may also explain why sharks might feast, consuming a large quantity of food, and then not feed again for several days or weeks.

The nutrients derived from the digestive process are used to power the shark's metabolism. As in most other animals, nutrients that are not needed immediately are stored for later use when food might not be readily available. The primary storage organ is the shark's liver, which reportedly can reach between 20% and 30% of the body weight of a healthy shark and may change seasonally depending on the scarcity of food sources.

Shark livers are also very high in vitamin A. Prior to the 1950s, when vitamin A was synthesized in the laboratory, there was a worldwide market for vitamin A purified from shark livers. The vitamin was added to milk to meet the daily human need for this vitamin. A wonderful book, *Shark! Shark!,* written by William E. Young and published in 1934, details his adventures as a commercial shark fisherman and the efforts to supply this oil for leading milk producers in the United States. Other compounds, such as squalene and squalamine, have been isolated from the livers and have shown some promise in medical applications.

Most animals' livers serve as energy-storage organs in addition to their other functions. But, unfortunately, they also store compounds from food sources that might not be desirable. Recent biochemical studies have shown the presence of many chemical pollutants in shark livers. Mercury, pesticide residues, and other contaminants might become a toxic challenge to sharks if their concentrations continue to increase. In addition, they may eventually interfere with extracting the valuable substances from shark livers that could be medically beneficial to humans.

Making Sense of the Senses

As top-level predators, sharks are undeniably well equipped for prey capture and feeding once they detect suitable food. How they are able to locate potential food sources has been the topic of study for decades, as scientists attempt to determine how effectively each of the shark's sensory systems functions.

Smelling Attractive

The sense of smell is considered the shark's greatest adaptation for locating potential sources of food. A sizable portion of the shark brain is devoted to olfaction (the sense of smell), but the exact sensitivity of the system is not well understood. The amount and type of stimulus, whether the scent of a prey item or blood or other attractants in the water, is influenced by the quantity of the odor present. It is also dependent on the depth of the water and associated water currents that dilute the odors. Small amounts of blood, for example, in very deep water with a fast current would be diluted in the very large dilution corridor, so detection may be unlikely. Conversely, a large volume of attractant in shallow water with a modest current may resist dispersal. Under these conditions sharks may be attracted from long distances as they detect the odors and swim up-current to the location.

Sharks detect substances dissolved in the water as the water flows into openings in the snout called nares, the equivalent of nostrils. Here odors are detected by the olfactory bulb, a chemical detection structure that may range from 3% to 12% of the shark's total brain mass. This system seems to respond well to attractants such as blood or other body fluids that may contain amino acids, the building blocks of proteins.

The nares are located on each side of the snout. As a shark attempts to locate a prey item it has detected, it often swims with its head moving side-to-side. It is believed

The external openings of the olfactory system, the sense of smell, are the nostril-like nares, located on the snout of the shark. In this extreme close-up of a tiger shark, the unique angular teeth can be seen in the front of the lower jaw.

that this motion exposes each of the nares to the odor corridor, helping to pinpoint the location of the prey. Some species, such as the nurse shark (*Ginglymostoma cirratum*), have sensory barbels, appendages that look a bit like whiskers. These structures can be dragged across the bottom with the side-to-side swimming movements and may aid in detecting food sources buried under the sand.

While it is important to understand the role of the sense of smell in detecting food, it may also be important in other functions as well. While very few studies have examined the production of sex attractants, called pheromones, those who have witnessed mating in the wild are convinced that smell most certainly plays a role. Watching a male shark detect and approach a female from a very long distance during the mating season, in dirty water with noisy surf, indicates that smell probably plays a very large role in the process.

Eye See You

The visual system in sharks has been studied from many different perspectives, but the results are often inconclusive. Their eyes are similar in structure to those of many other vertebrates, and their visual acuity is similar to that of other fish. The eyes of sharks possess several different types of cells. They have rod cells, which are generally associated with seeing in black and white and grayscale. Their eyes also seem particularly well suited for vision under low light conditions, such

as dawn and dusk, times that are often correlated with shark hunting and feeding behaviors. This capability is enhanced by a crystalline layer located behind the retina. This layer, called the tapetum lucidum, is present in many nocturnal animals. It is thought to function by reflecting light entering the eye until it actually strikes a light-sensitive cell and is thereby detected.

Cone cells, most often associated with color vision, are also present, although their presence, type, and density vary from species to species. Whether sharks can detect color has been a matter of some controversy for many years. It is difficult to tell whether a shark is detecting a particular color or responding instead to different light levels or to varying degrees of contrast. It is also generally accepted that color vision requires several different types of cone cells, each responding to a particular color. Most species of shark have only a single cone type, thus suggesting that color vision is either absent or limited. Some skates and rays, by contrast, have several types of cone cells and are considered highly likely to be able to detect different colors. Behavioral studies to examine color vision have not been wholly conclusive, and the question of shark vision remains an active topic for research. Since sharks inhabit waters that are often deeper than 40 to 60 feet, the ability to detect color would not be a particular advantage; most of the colors that make up white light are filtered from the water at such depths. Divers know that reds, oranges, and yellows often disappear quickly as they descend. Photographers, in fact, must supply their own light with the use of flash attachments or video lights to detect these colors for photographs or videos.

Can You Hear Me Now?

Detection of sound by sharks relies on their ability to detect the movement of water particles that are displaced or caused to vibrate by a sound source. These vibrations are sensed by an inner ear containing sensory

One of the diagnostic traits of the nurse shark is the strange sensory barbels that are found on the snout. They are thought to aid in detecting the chemical signatures from prey items that may be buried under the sand or mud.

hair cells that can respond to vibrations originating from a source of sound.

Some rather novel experiments were used to establish what sounds were particularly attractive to sharks. A group of young scientists happened to be spearing fish one afternoon and noted that wounded fish emitted a sound that was detected by the divers as well as local sharks. They returned another time with underwater microphones to record these sounds. Afterward, in the lab, they used artificial sound generators to duplicate the sounds. Returning to the reef for another dive, they lowered an underwater speaker to an area where they suspected sharks could be found. They played back the actual recording of a struggling fish, and then played their artificial recording. They found that both attracted the sharks.

Analyzing the recordings, they determined that the most attractive sounds were at very low frequencies. These are typically considered to be the same frequencies of sounds emitted by struggling or wounded fish, or struggling swimmers. While it is not as sensitive as the hearing of bony fish, in which the air-filled swim bladder may assist in sound detection, shark hearing is particularly sensitive in the frequency ranges that are most important in the shark's world for locating its preferred food sources. Sharks' ability to detect and localize such underwater noises is a further refinement in their sensory adaptations for effective predation.

Aiding the inner ear in detecting vibrations is the shark's lateral line system. This is a series of pores in the snout and along the sides of sharks and bony fish. Vibrations detected by sensory hairs within these pores can aid in finding a target worth investigating. If a shark is presented with a sound on its right side, sensory cells on that side will detect the sound while those on the left will not. In this way a shark can turn toward the source, activating even more sensory cells on the snout. The shark then determines the best direction to swim to encounter the

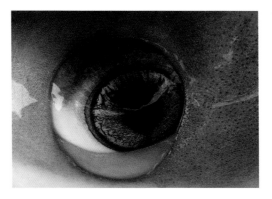

The eye of a tiger shark is remarkably well suited to this shark, a predator at dawn and dusk.

sound source if it interprets it to be worth investigating. These same cells can also respond to currents and changes in currents, helping the shark decide which direction to move.

Electrifying Sensations

One of the most useful and unusual shark sensory systems is the electrosensory or electroreception capability. Sharks possess very highly specialized receptors called the ampullae of Lorenzini that develop from the lateral line system. These structures are pores containing a gel-like substance that is especially sensitive to weak electrical currents. All living organisms generate small electrical potentials whenever their nerves fire and their muscles contract. Since these currents are very small, they cannot be detected over long ranges. Most sensory biologists believe that sharks are able to utilize this sensory capability when they close in on a living prey item and that it guides the shark during its final and closest approach. At this time the jaw may be open and the eyes may have lost contact with the food source. Having a sensory system that functions over very short distances may reduce the chances that a food item can successfully escape.

The early investigators of this system conducted some unusual experiments that were fascinating in their simplicity and elegance. Using a small species of shark kept in small pools with a sand bottom, they decided to test the various sensory systems for their role in predation. They first provided small living fish as a food source to the experimental group of captive sharks. The sharks fed as expected. The next part of the experiment tested the role of vision. Because of the structure of the sharks' eyeballs, the scientists were able to cover the eyes with a piece of plastic and eliminate the sharks' ability to see the food fish. The sharks were still able to find the fish and feed on them. Next the sharks were further deprived of

The small pores on the snout of this tiger shark are the ampullae of Lorenzini. They are gel-filled structures that are part of the electrosensory system of sharks and rays that aid in detection of prey from short distances.

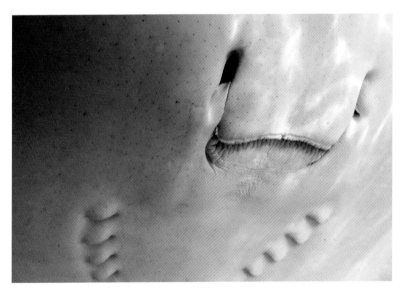

The ampullae of Lorenzini are very visible in the snout of this southern ray. Rays depend on this system to detect the weak electric currents from fish or other prey items that may be buried under the sand and not visible to the ray.

sensory input by an anesthesia inserted into the sharks' nares, eliminating their sense of smell. The sharks still found the food, but not as quickly. Reasoning that the sharks were probably detecting vibrations from the fish, the investigators built a small chamber from agar, a gel that has no real odor properties of its own but prevents vibrations and smells from escaping. They then enclosed the fish in the agar chamber and buried it under the sand. At this point, three sensory systems were nonfunctional. Yet the sharks could still find and devour the fish. When a dead fish was placed in the agar, however, the sensory-deprived sharks were not able to find their food. Knowing that the only difference between a living fish and a freshly dead fish was the electrical currents emitted from the live fish, the investigators replaced the dead fish with a set of electrodes that emitted a current identical to that of a live fish. The sharks ate the electrodes. It took it longer to find them, and the sharks were able to detect their presence only when they swam directly above the electrodes, illustrating that the electrosensory mechanism works best at very short distance.

The electrosensory system is also thought to be an important aid in guiding migratory movements. Since the earth produces a magnetic field, receptors that can detect this magnetism may help animals to orient along these lines and enhance sharks' ability to move to specific areas as they migrate.

A common question about shark senses is which sensory system is most effective and which one the shark most relies on. There is really no good answer, because it depends totally on how strong the stimulus might be. It's difficult to determine which would be more attractive, 55 gallons of blood in a huge ocean, or explosions from a boat fire, for example. To properly test which system is most important would require creating a test in which all stimuli are somehow equivalent. That experiment has yet to be conducted. It is likely that sound or smell works at the longest distances, again depending on how loud the sound might be and how much scent is available from the source through the dilution of the currents. Vision probably works at shorter distances, depending on the clarity of the water. Electroreception is likely to function in the last few feet of a feeding encounter.

Understanding exactly how shark sensory systems function is the first step in developing an effective shark repellent. Studies during the Second World War to aid sailors and fliers who might have been shipwrecked or downed at sea centered on developing repellents based on chemical detection. Many substances were found to be ineffective, but a compound of copper acetate had some success. This was later combined with a dye that surrounded swimmers, theoretically hiding them from nearby sharks. Called Shark Chaser, the packets were distributed to fliers and sailors in the hope of easing their fears of being cast adrift in waters containing sharks. This attempt probably gave a sense of well-being to some service members, but later studies showed the repellent to have very little effectiveness. A variety of other substances have been tried over the years, including toxins from the Moses sole (*Pardachirus* sp.) and various detergents. But their effectiveness as repellents has not been as promising as many would hope.

Confusing the shark's eyesight to serve as a repellent has also been studied. The dyes in Shark Chaser had offered some clues that vision could be confused. Creating screens of bubbles surrounding beach areas where sharks are present has been tested with little success. The Johnson Shark Screen was a large plastic bag with an inflatable collar, into which a person could climb. The bag had no particular shape, so it disguised the form of a body well. And if the person was bleeding, the fluids would be contained within the bag, so presumably no scent clues would be present for the shark to detect. These remain in use and have spawned a market for the development of other ways to camouflage a per-

Shark Chaser, one of the first chemical shark repellents developed during the Second World War, was little more than a psychological boost to fliers and sailors. It was rushed into production to offer some form of protection from shark attacks at sea, but later studies showed that it was not a very effective deterrent.

son in the water by designing what are called cryptic wetsuits. These are designed to confuse the shark's vision, to make the swimmer or diver look like nothing or like something with no appeal to a shark. Testing continues.

There have also been attempts to discover whether some particular sounds might discourage or even repel sharks. If certain sounds attract sharks, it seems reasonable to assume that some sounds may repel them. While this seems logical, if sharks cannot detect sounds above a certain level, then perhaps sounds outside this region are simply not detected. Some natural sounds, such as the loud noises of killer whales (*Orca* sp.) have been shown to repel lemon sharks. The absence of killer whales in appreciable numbers in Florida and Bahamian waters makes one wonder exactly what characteristics of these sounds are disturbing to tropical sharks.

Scientific studies have recently directed more attention to the electrosensory system. In field tests, small electrical currents emitted from batteries attached to hooks have shown some promise in discouraging sharks from taking baits. Some adaptations have been made to create diver-wearable devices that emit a current that repels rather than attracts sharks; such devices are available on the commercial market. Additional studies have shown that various types of magnets have a repulsive effect, presumably by overstimulating the shark's electroreception capability.

It seems in all these cases that a better understanding of sensory systems, especially how they may be confused or overstimulated, may hold the keys to keeping sharks away when they are not wanted. How these various signals are interpreted by a shark's brain and how a response is evoked is an area of intense research. Modern techniques have been developed that permit electrodes to be implanted in the brain to determine what kinds of stimuli evoke a response from a shark. In this way, determining what attracts and what repels a shark may eventually be discovered, especially when studying factors that result in attacks on humans by sharks.

When Is an Attack Not an Attack?

It's probably fair to say that the reason people know about sharks, are concerned about sharks, and even fear sharks comes from attacks on humans. While there are definitely attacks, and people have been killed by sharks, not every encounter is an attack and seldom do human–shark interactions result in injuries. The International Shark Attack File, maintained by the Florida Museum of Natural History, compiles all the information they can find about every reported shark attack around the world. The file has had many homes in the past, including the Office of Naval Research. Data were then passed along to the Smithsonian Institution before eventually finding their permanent home in Florida.

Analysis of the data in the file indicates that fewer than 75 unprovoked attacks occurred worldwide in the most recent analysis, with only three fatalities! It is seldom the case that more than 100 attacks, world-

wide, are recorded. People are fond of comparisons to events they can relate to, and it has been suggested that the chances of being attacked by a shark are less than being struck by lightning (38 deaths per year in coastal states) or being bitten by other people in New York City (10 times more likely than being bitten by a shark, worldwide!).

The entertainment industry and the news media are the primary culprits in creating the sharks' bad reputation. Photographs of sharks swimming along behind a kayak, with no encounter, no bites, no bumps of the kayak, are still accompanied in media accounts with the headline or caption "Shark Attacks Kayak." Viewers anxiously await new episodes of shark "documentaries" that reveal myths and legends cultivated by such shows in quest of large audiences. Truth seldom gets in the way of entertainment!

A suggestion of more appropriate terminology has been developed to counter these sensationalist headlines. It is hoped that this phrasing will be widely accepted and adopted as a way to more clearly identify and classify shark–human interactions. From a paper written by Christopher Neff and Robert Hueter:

1. Shark sightings: Sightings of sharks in the water in proximity to people. No physical human–shark contact takes place.
2. Shark encounters: Human–shark interactions in which physical contact occurs between a shark and a person, or an inanimate object holding that person, and no injury takes place. For example, shark bites on surfboards, kayaks, and boats would be classified under this label. In some cases, this might include close calls; a shark physically "bumping" a swimmer without biting would be labeled a shark encounter, not a shark attack. A minor abrasion on the person's skin might occur as a result of contact with the rough skin of the shark.

3. Shark bites: Incidents where sharks bite people resulting in minor to moderate injuries. Small or large sharks might be involved, but typically, a single, nonfatal bite occurs. If more than one bite occurs, injuries might be serious. Under this category, the term "shark attack" should never be used unless the motivation and intent of the animal—such as predation or defense—are clearly established by qualified experts. Since that is rarely the case, these incidents should be treated as cases of shark "bites" rather than shark "attacks."

4. Fatal shark bites: Human–shark conflicts in which serious injuries take place as a result of one or more bites on a person, causing a significant loss of blood and/or body tissue and a fatal outcome. Again, we strongly caution against using the term "shark attack" unless the motivation and intent of the shark are clearly established by experts, which is rarely the case. Until new scientific information appears that better explains the physical, chemical, and biological triggers leading sharks to bite humans, we recommend that the term "shark attack" be avoided by scientists, government officials, the media, and the public in almost all incidences of human–shark interaction. (Neff and Hueter 2013, 70)

Dating and Mating

Of all aspects of shark biology, the one we understand the least from field studies is shark reproduction. Studies of dead animals caught by commercial fishermen or by recreational anglers have given biologists hints about it. Yet the insight gained from actual successful field observations may be the only way to understand what conditions are necessary for shark populations to maintain stable levels. This is especially true for the species of sharks that have some commercial value. Without a complete understanding of the rate at which commercially valuable

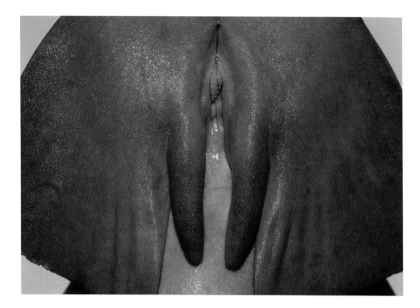

In male sharks the pelvic fins are modified to form the claspers, the tubular structures shown here in a juvenile nurse shark.

In an adult male shark, the claspers are more prominent and make identification of the sexes easier.

species replenish their numbers, no sound management practices can be developed that consider both the well-being of natural populations and the welfare of fishers who depend on constant numbers of sharks for their living.

Studies of the internal and external anatomy reveal that sharks mate like many other vertebrates. The male has modifications to the pelvic fins called claspers. These are penislike organs that are inserted into a female during a mating event and through which sperm is passed from the male into the female. Only one clasper is used in a mating event.

Females have two uteri, where the eggs or the young will develop after fertilization has occurred. In some species there is a shell gland through which eggs pass as they move from the ovary through the reproductive tract. Fertilization must occur before the egg case is created, or no embryo will develop, since sperm will be unable to penetrate the egg case. Some species incubate their fertilized and developing young in both uteri, while other species may use only one uterus.

The oceans occupy a huge portion of the earth's surface, so finding a mate can be a challenge. Several shark species have developed novel ways to get around this problem. The female simply stores the sperm for long periods until she is physiologically and developmentally prepared for a pregnancy. At that time capsules that may have surrounded the sperm are digested away and the sperm is released. Fertilization may then occur, and it may be many years after mating has occurred. For example, a female kept in an aquarium for 4 years without any males eventually gave birth to healthy offspring. DNA analysis showed that the pups did have a father, so mating had to occur sometime before the female's capture. This is an extreme example but shows the marvelous reproductive adaptations that sharks possess.

Equally curious is the case of several sharks that have given birth without the aid of a male. This process, called parthenogenesis, is known to occur in some fish, amphibians, and reptiles. It has only recently been described for some species of sharks, including both the bonnethead shark (*Sphyrna tiburo*) and the blacktip shark (*Carcharhinus limbatus*), both of which are common to Florida and Bahamian waters. This type of birth has also been confirmed in the spotted eagle ray (*Aetobatus narinari*), another common Florida and Bahamian resident.

I Think We're Pregnant

Once the eggs have been fertilized, several developmental strategies, which have been observed in different shark species, allow the embryos to mature. It seems that sharks and their relatives have tried every form of development possible in their 400 million years of existence.

In some species, the fertilized eggs are shed soon after they have been encapsulated by the shell gland. The eggs may simply fall to the ocean floor or become entangled in some growth on the bottom to await completion of the embryos' development and eventual hatching. These egg-laying species are said to be oviparous ("ovi-" referring to eggs). Skates are examples of oviparous species.

Another mechanism is for the fertilized egg to develop within the mother. The young derive nutrition from the yolk of the egg until their development has been completed. The young hatch inside the mother, the egg cases are expelled, and soon thereafter the young are born. During the final stages of embryonic growth, the young may actually be nourished by secretions produced by the uterine lining of the mother where the young are incubated. These secretions, often referred to as uterine milk, or matrotroph, sustain the animals until birth. This form of embryonic development occurs in nurse sharks. Because the young are born live, and because there is no placenta attaching them to the mother, this type of reproduction is termed aplacental viviparity.

The final method involves the formation of a placental attachment between the developing young and the mother, similar to the development of human embryos. When the nourishment from the yolk sac is depleted, the sac develops as a connection to the mother that provides nourishment through the embryonic stage of development. When development is complete, the young are born,

An endoscope equipped with lighting and a camera is used to photograph a newly hatched nurse shark embryo inside the mother. In this form of development, eggs mature within the female until they hatch. The empty egg cases are then shed, and several weeks later the young are born alive. This procedure was one of the first uses of endoscopy to evaluate shark pregnancy in living sharks. It was conducted in the surgical suites of Sea World in Orlando with veterinarians Dr. Sam Dover and Dr. Mike Walsh.

often with small remnants of the placental cord still attached. Most of our sharks in Florida and the Bahamas develop in this way. It is known as placental viviparity to distinguish it from aplacental viviparity, in which embryos lack a placenta. Once the young are born, there is no parental care. The newborn shark pups, or neonates, are on their own from the day they are born.

The number of young in a shark litter is variable, differing from species to species. Some larger species may have only one or two, perhaps because nourishing multiple large embryos would require more maternal resources than are available. Nurse sharks have as many as 30 or more in a litter. Whale sharks, not surprisingly because of their huge size, have had litters with as many as 300 embryos.

One unusual feature of some species of developing sharks is their taste for littermates. Cannibalism within the uterus has been shown in several species, but most notably in the sandtiger shark (*Carcharias taurus*). Consuming younger, smaller littermates may allow an embryo to grow to a relatively large size in the uterus and thus be better prepared against other predators when it is born. Some species may consume unfertilized eggs as they develop, an action called oophagy (oocytes are immature egg cells; "-phagy" means eating a specific type of substance; oophagy therefore refers to consuming the unfertilized eggs). Both cannibalism and oophagy are further examples of the strange and different reproduction mechanisms in sharks.

The first hints of how sharks actually mate was the discovery that at certain times of the year, many shark females are found to have severe bite marks or scars, mostly on their pectoral fins or around the head. It wasn't until observation was done in the field and in captive facilities that the origin of these scars became better understood. Since shark mating involves insertion of a male's clasper into the female, a male must find some way to grasp the female and position himself in

a way that allows the clasper to be inserted. The mechanics of these movements are difficult.

Some aquarium studies with smaller sharks have shown how it occurs. But larger species have been reluctant to mate in captivity, so few records of mating and successful birth in captivity exist. Observations in the field environment have been equally difficult, requiring knowledge about where reproducing populations are and when mating is likely to occur, and then observing sharks actually mating in waters that may be very deep. The only species in which mating sharks have been studied in a systematic way is the nurse shark. This species chooses shallow, near-shore waters where the animals can easily be observed, and we have taken advantage of that to study their mating behaviors.

My research colleague, Wes Pratt, and I began a study of mating nurse sharks in the Dry Tortugas in the early 1990s. For more than 15 years we tagged, measured, observed, and filmed these animals through more than 1,000 mating attempts. We found that not every attempt was successful and that unless females selected a suitable mate, whatever that might mean to a shark, mating was unlikely to occur.

Females employ a variety of choice and escape maneuvers to select a mate. Once a mate has been chosen, the delicate dance begins. The male will bite the female's pectoral fin

Nurse sharks' use of shallow water as mating grounds makes the events of mating easier to recognize and study. Approaching the animals during mating events seems to be easier when the animals are distracted.

and position himself so that the clasper is inserted and the mating begins. It may last 2 to 10 minutes, and afterward the pair separates. A female nurse shark will mate with more than one male during the period in which she is reproductively active. This period may last for 2 to 3 weeks and then not occur again until 2 or more years later. Other species of sharks are thought to have similar reproduc-

The establishing of an adequate grip to position the sharks for actual mating involves a series of delicate maneuvers. A male grips the female's large pectoral fins and then positions himself so that one of the claspers can be inserted and copulation can begin. In this photograph, the entire pectoral fin of a female is in the male's mouth as he grips the female to prepare for mating.

Once a clasper has been inserted into the female, copulation begins, and sperm will be introduced into the female to fertilize her eggs.

tive strategies, though no species has been observed and studied more closely than nurse sharks.

They Look Just Like You . . . and You . . . and You!

In the years of our study, we were able to capture several females after mating. To ensure that they were carrying eggs and were therefore possibly pregnant, we conducted ultrasound examinations aboard our research vessels with the help of veterinarians from SeaWorld. Suitable females, those that were carrying eggs and could have been pregnant, were then taken to SeaWorld Adventure Park in Orlando, Florida, where they were separated and kept in nondisplay tanks throughout their pregnancy. These studies revealed that the females were pregnant for 4.5 to 5.0 months.

Once the pups were born, the DNA analysis of the young sharks revealed that a litter of 24 to 30 nurse sharks might have between five and seven fathers, one of the first examples of multiple paternity in sharks. This is now known to occur in many other shark species, and it is a successful way to make certain that populations of sharks maintain some genetic diversity, especially for those species that don't migrate over long distances.

Getting along with Others: Symbiosis

Sharks share their realm with remoras, perhaps the ocean's most common freeloader. With its dorsal fin modified as a suckerlike structure, a remora is able to hitchhike on the back of a shark and enjoy the leftovers from a shark's meal. Often called shark suckers, these fish may not really qualify as freeloaders. Recent videos and photographs have shown that they perform a valuable function for sharks by removing external parasites. Sharks are apparently somehow aware of this benefit and will even allow remoras to swim into their open mouths to clean away parasites there. This is very similar to the role played by many cleaner fish on coral reef systems that perform similar functions for bony fish. In addition to remoras, pilot fish are often seen to accompany sharks, companions that are thought to perform some of the same cleaning functions as remoras. Pilot fish also seem to be recognized by sharks as beneficial and may thereby escape predation.

Hazards to Their Health

Take Two Sardines and Call Me in the Morning: Diseases and the Immune System

It is commonly believed that sharks are immune to disease, including cancer. This is not true, but the shark immune system is very advanced—though not on the same levels as for mammals—and sharks seem to be naturally protected from many bacterial diseases that may infect other fish. They possess natural immunities but are also able to cope with inflammation and many types of infections that they are not naturally protected against. Nevertheless, they can be adversely affected and can even die from some viral infections. We often encounter sharks with very large, open wounds, many of which occur during mating. While their skin is certainly tough and resistant to many types of scrapes and violent feeding activities, injuries do occur and can lead to serious infections. Sometimes the rubbing that occurs from tags will leave tissue exposed and, because the rubbing from the tag is continuous, these open sores can be slow to heal. The healing process is usually more rapid, however, and it is often difficult to find evidence of scar tissue when healing is complete.

A large body of scientific literature describing sharks' resistance to cancer has been produced. Because of the mistaken belief that sharks have a natural immunity to cancers, natural products from sharks have been marketed that purport to prevent or treat human cancers. However, sharks with large tumors have been captured or taken by commercial fishermen. These diseased animals show that

Remoras are nature's best hitchhikers, and sharks are seldom found without one or more attached to their skin. The remora's dorsal fin is modified to serve as a suckerlike attachment; remoras are able to hold on during even the most vigorous exercises. It is not uncommon for a remora to seek out a diver and tag along for the ride. The experience is a bit unsettling but not a danger.

The lemon shark in this photograph may represent an underwater remora condominium, providing accommodations for an entire school of remoras.

sharks are not totally immune to cancers. In addition, some cancers in captive sharks, including severe forms such as melanomas, have been reported by veterinarians. One possible reason we do not see many tumors in sharks is that if they are severe or fatal, the

Sharks such as this oceanic whitetip shark are often seen in the company of one or more pilot fish. Their role as cleaner fish may provide a symbiotic benefit to the sharks they accompany. Once thought to lead larger animals to their food, pilot fish are actually freeloaders that gain protection and access to scraps by associating with large sharks. They also save energy by riding the pressure wave created by the shark's forward momentum.

The mating dance requires that the male bite the female pectoral fin to establish and hold his position during mating. There are often scars from the male's teeth that could potentially injure the female. The skin of the female is actually thicker than that of the male, affording some protection from the bites received during mating. Fresh bite marks can be seen in this grainy picture of a female after a mating event. The water is often stirred up during the event, so clear photography is not always possible

animals probably die and therefore are not captured. Thus their tumors are never discovered.

Even though sharks are considered to be among the top predators in the world's oceans, they are not without natural enemies. Other sharks, large groupers, and toothed whales such as *Orcas* often feed on sharks, especially smaller species. It is not uncommon when one is cleaning and filleting a large grouper to find a shark or shark remains in its stomach.

Sharks are also often plagued with smaller parasitic pests that can seem like enemy invaders. Studies have identified more than 1,500 different parasites, ranging from leeches and lice to roundworms, flatworms, tapeworms, and more exotic-sounding animals such as isopods, copepods, ostracods, and flukes. While no one would likely be surprised that shark digestive tracts have parasites, almost every other tissue in a shark can serve as a home for parasites as well. Different species of parasites have been identified from the shark digestive tract, the gills, the skin, the eyeballs, the gall bladder, the heart, and the blood vessels. External para-sites can often be seen on captured or free-swimming animals. Copepods, small crus-taceans, are very common, and occasionally leeches can be found still attached to shark skin, gills, and the insides of their mouths. In fact, parasitologists believe that no shark tissue is likely to be immune to all forms of parasite. Though parasites can be a problem, parasitic infections are generally not fatal. A "good" parasite is one that causes the least amount of damage or harm to its host. It does not serve a parasite well if it kills its host and must then find another.

Pollutants

An emerging concern is the detection of pol-lutants in shark tissues. Mercury, pesticides, and perhaps even by-products from oil spills may be making their way into sharks' tissues at potentially dangerous levels and may be a more silent enemy. Sharks have a remarkable ability to adapt through time to changes in their environments. But between toxic chem-icals from human activities and the gradual changes in oceanic temperatures, there is some concern that their adaptive abilities may be encountering threats from which they cannot recover.

Even Sharks Need Less Stress

Injuries, diseases, and parasites can possibly disrupt the mechanisms that maintain nor-mal health and physiological stability for any animal—that condition that biologists term homeostasis. Any challenge to homeostasis can be categorized as a form of stress. How sharks respond to stress, though of great concern, is a relatively unstudied area. With many populations of sharks under threat from overfishing, more and more studies of shark age, growth, and movements have been undertaken. These studies rely on capturing sharks, perhaps even removing them from the water for short durations, so that mea-surements can be taken and tags and tracking devices attached. Sharks are strong fighters

when captured, and studies have shown that biochemical changes occur immediately and may persist for some time after capture. Changes in circulating blood sugar levels, electrolytes, and some hormonal changes may have longer-term effects than are cur-rently understood. Scientists have some con-cerns that when sharks are lifted from the water and lack the buoyancy provided by the

Marine leeches are also commonly found on sharks that inhabit near-shore waters. Leeches have mouthparts that allow them to attach to even the toughest skin, including that of sharks. It is not known whether the leeches are able to penetrate the thick skin to derive any nutrition from the shark or whether they must move to softer areas such as the mouth. Most shark parasites are considered pests but are usually not life threatening.

Sharks are very fre-quently found with many external parasites, such as the copepods that adorn the dorsal fin of this mako shark. Copepods are also commonly found on the eyes of sharks but apparently do not compromise the shark's vision. Definitive studies have not yet been con-ducted.

Working on a shark to obtain measurements and attach tags requires care and attention to the animal's stress levels while protecting the scientific staff. Providing a flow of water across the gills and supporting the animal's weight are important steps to ensure that the animal will be released in good condition. This large lemon shark was measured, tagged, and released in just a few minutes and actively swam away when it was turned loose. *Photograph courtesy Alan Chung. Used with permission.*

water, the weight of the internal organs might place the animals at some risk for damage to these organs and the thin tissues that support them. Furthermore, a current of water must be artificially directed across the gills, usually from a seawater hose placed in the mouth, to keep the animal from suffocating. Careful handling of sharks, whether by anglers or by scientists, is crucial to minimizing the effects of stress and ensuring the long-term health of captured animals.

On Their Best Behaviors

In spite of their reputation as predatory killers, sharks are actually very complicated animals with a range of behaviors that may surprise most people. Attacks on animals or people and frenzied feeding are often the first responses given when people are asked what behaviors they think best characterize sharks. Feeding frenzies, the group behavior of large numbers of sharks wildly competing for bait or food, has been portrayed as a common occurrence. Reports of sharks attacking other sharks and of injured sharks continuing

to feed, even with large open wounds and their entrails streaming out of their body cavity, fuel the almost mythical fear of sharks. In fact, these two behaviors represent only a subset of more complicated actions that reveal capabilities more often associated with animals considered more advanced than sharks.

For example, more than 50 years ago, Eugenie Clark, one of the world's foremost experts on sharks, conducted classical conditioning experiments to test the abilities of lemon sharks to learn to press a bell for a food reward. The experiments were conducted at what was then known as the Cape Haze Marine Laboratory in Sarasota, Florida, now named Mote Marine Laboratory. The lemon sharks eventually learned to associate the food reward with pressing a target that caused a bell to ring. The experiments were discontinued when water temperatures dropped, and then were restarted after a 10-week delay. What is more remarkable than their ability to learn was that the sharks retained their training during this interval and once again pressed the target for their reward when the experiments were resumed, suggesting the existence of memory.

Responding to a source of food is not surprising to fishers who clean their catch, often with crowds of nurse sharks huddling under a cleaning station. Divers who participate in shark eco-encounters are able to observe sharks who appear at the approach of a boat carrying frozen fish remains (affectionately called chumsickles) because of their apparent ability to associate the boat sounds with an imminent food source. Far from the legendary feeding frenzies of "sharks gone crazy" over a food source, the animals are generally mild-mannered and easy targets for photography. Species that would normally be avoided in the water, such as lemon sharks, bull sharks, tiger sharks, and very large hammerheads, are common participants and seldom pose a threat to divers, though the Bahamas caution against feeding bull sharks.

It is important to note that the dives yielding these observations are highly structured and choreographed and rely on an associative behavior that is conditioned over a long period of feeding encounters. Simply jumping in the water on a favorite reef with a box of chum is not recommended; it is risky behavior, at best.

Wanderers, Vagrants, and Homebodies

Shark movement patterns, whether short-distance movements or long-distance movements, are the object of intense investigation. Animals must be caught and marked in some way for their movements to be tracked. Since 1962 sharks have been tagged by the Apex Predator Investigation, a project of the National Marine Fisheries Service (NMFS), which is part of the National Oceanic and Atmospheric Administration (NOAA). NMFS biologists, other scientists, recreational anglers, and commercial fishermen have tagged more than 230,000 sharks, representing 52 different species. These initial studies used what was called a spaghetti or dart tag. Other tags, even including cattle ear tags, have been used by other investigators.

When sharks are tagged with NMFS tags, measurements of their length, their weight (when it can be measured), the shark's sex, and the location where the shark was tagged are recorded and the information transmitted to Narragansett, Rhode Island, where the investigation is headquartered. When a tagged shark is recaptured and similar data are recorded, biologists are able to determine the movement of the shark and its rate of growth. The NOAA website reveals that some species show no or little movement and that others may have moved as many as 3,997 nautical miles (4,596 statute miles). Animals have been recaptured at intervals as long as 27.8 years, this in the case of a sandbar shark (*Carcharhinus plumbeus*) tagged in June 1965 and recaptured in March 1993.

These data can also provide hints about the life expectancy of sharks. The commonly cited possible life span for several common species is generally considered to be 40 to 50 years, but some studies have shown that the Greenland shark (*Somniosus microcephalus*) may live nearly 400 years. These sharks are thought to have the longest life span of any vertebrate species.

Not all species of sharks move across the ocean, as blue sharks (*Prionace glauca*) do, and not all species move from north to south across the equator. Some stay within a small

Perhaps the best known shark behavior to almost everyone is frenzy feeding. This is a seemingly uncontrolled feeding activity that occurs with large groups of sharks and blood or food in the water.

Coordinated shark dives have become popular and provide divers with an opportunity to see and dive with sharks in very close encounters. Dive trips are highly structured, and incidents with sharks are rare. Dives that place bait in the water to attract the sharks are not permitted in Florida waters.

area and are considered residents, including the common nurse sharks of Florida and the Bahamas. While they may indeed show some movements, the movements are not on a scale comparable to those of blue sharks or sandbar sharks. Other species may leave for short periods, even traveling long distances, but return to the same area, often seasonally, perhaps for feeding or, in the case of the nurse shark, for mating. This periodic and faithful return to the same area is called site fidelity. Some species return to the specific area of their birth to give birth to their own young; they are said to be philopatric. A familiar example of such behavior is the return of salmon to their home streams to lay their eggs. Tagging and tracking studies have confirmed that some species of sharks, such as the lemon sharks and nurse sharks in Florida and the Bahamas, display this same behavior.

While tagging studies are vital to understanding shark age, growth, and movements, recapturing a shark shows only where it was at the time of capture and recapture, unless the animal is released and recaptured many times. New technological advances permit a more constant record of movements. Using ultrasonic transmitters that are attached to or implanted within a shark, its movements may be detected by special receivers placed on the ocean floor. Radio waves are not transmitted in water, so ultrasound must be used instead. Each transmitter is uniquely coded so that individual animals can be identified. These transmitters do not transmit over long distances, so many monitors must be used in areas where sharks with the transmitters are likely to spend time. Cooperation between researchers using this technology is growing. Scientists in one area studying a certain species may detect different animals of a completely different species. These data can be exchanged and greatly broaden the area over which movements can be detected.

A more recent advance has been the use of

satellite telemetry. Special tags that communicate with orbiting satellites are capable of tracking sharks over long distances for long durations and can give a continuous record of where individual sharks are at any given moment. The telemetry only works, however, when the shark's dorsal fin breaks the surface. Because of this limitation, the technology is less useful for bottom-dwelling species. The system that uses the satellite tags and the satellite time is enormously expensive but has the promise to reveal patterns never before available.

One of the best known of these studies is the work of OCEARCH, a not-for-profit organization that cooperates with scientists to tag sharks with this form of monitoring and makes the tracking data available in real time to anyone interested in following the movements of named animals that carry the tags. This open-access approach to studying shark movements has a great appeal to scientists and amateurs alike, and particular sharks, such as "Katharine" and "Mary Lee," have almost cultlike followings.

Another, very different way to gain insight into the lives and behaviors of sharks is to record their every activity. For many years the National Geographic Society has been developing and refining the use of Critter-Cam, a video imaging system that attaches to an animal, records its activities, and then is released from the animal for later recovery. This technology has provided new insights into the secret lives of many marine animals, including many species of sharks.

Although it is useful to know where sharks are at any given time, none of these technologies explain *why* they move as they do. Until we can learn what is driving the movements of sharks, the story will remain incomplete. It is generally agreed that, just as in other species, movements may occur seasonally to follow changes in water temperature or movements of the sharks' preferred foods, or may occur for purposes of seeking a mate or to give birth. Detecting overlapping movement

patterns from satellite tracking may allow biologists to predict where the animals might gather, so that closer and in-person observations can be made of what might be bringing the animals to certain very specific locations. This development could help resolve the mysteries of the travels of sharks.

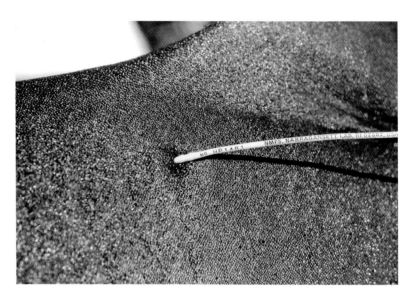

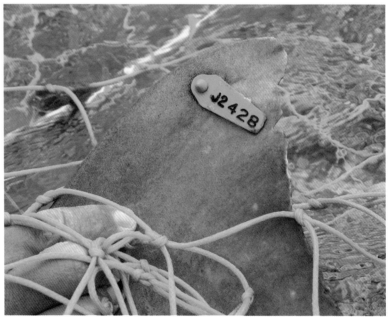

One of the earliest types of tag was the dart tag (*top*), often called a spaghetti tag. The tag asks a fisher who might catch a tagged shark for specific information and provides an address or contact information for a finder to respond to. A more visible tag was borrowed from cattle ranchers. The RotoTag (*bottom*) has a larger number on one side and allows better identification of individuals if the tag number can be read. This is most useful for sharks that stay in one location for longer intervals, when identifying individuals is important.

To track the movements of sharks, ultrasonic transmitters are often used. Because they do not transmit over long distances, many special monitors that can detect the transmitters are required. Each transmitter has a unique code, to facilitate identification of tagged individuals.

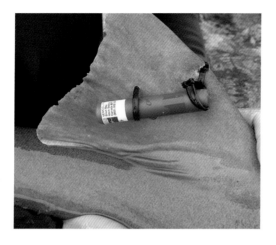

The most useful tag for monitoring long-time and long-distance movements is the satellite tag. Once attached to the dorsal fin of the shark, the GPS-equipped tags are capable of storing location and depth information and transmitting it to satellites. A limitation is that the antenna of the tag must be above the surface for a short time to allow the satellite to transmit its data. These tags do not work well for animals that never come to the surface. The technology is advancing rapidly and promises to reveal details about shark movements and life history never before available *Photograph courtesy Derek Burkholder. Used with permission.*

What Good Are Sharks?

Role in Aquatic Ecosystems

Every ecosystem on the planet, whether it is a terrestrial system such as a forest, a prairie, or a desert, or an aquatic system such as a lake, a pond, a river, or an ocean, shares a common organization termed a food chain, a food web, or a food pyramid. Such a structure describes how the organisms in the ecosystem relate to one another in terms of food and prey and how energy flows through the ecosystem.

Most depictions of these interrelationships place sharks at the top of the oceanic food chain as apex predators. This placement does not apply to every location and ecosystem; sharks may simply rank as high-level predators in some systems. A greatly simplified food pyramid rests upon the largest and arguably most important level, that of the primary producers. They form the base of the food pyramid. These organisms are the plants that convert radiant energy from sunlight into plant biomass, serving as the primary food source for the next level, the herbivores that feed on plant material. Carnivores and omnivores, collectively called consumers, make up the next level and feed on animals within and below that category. The top level consists of animals that feed on all levels below them. These are the apex predators; in many aquatic ecosystems they are represented by sharks as well as other large fish and many marine mammals. The size of each level of a food chain may vary between systems, depending on the variety and biomass of the organisms in the system. Each successively higher level must be smaller than the level below if the system is to remain in balance. The number of levels and the size of the levels may vary, but the general structure remains essentially the same. When a system is in balance, lower levels are kept in check by the levels above. A healthy level of herbivores prevents overgrowth of plant life, and the system remains healthy. Carnivores keep the herbivores in check. The same logic can be applied to every level of a particular food web. A healthy food web has a proper balance at all levels.

Biologists become concerned when one or more levels of a food chain are disrupted by elimination of one or more other levels or by overpopulation in any level. In the simple food chain example in the preceding paragraph, the disappearance of herbivores could result in overgrowth of the plants they would normally consume. And the predators that would normally prey upon these herbivores must now find another food source, perhaps preying on items they might otherwise ignore, challenging another predator's survival. If the apex predator is eliminated, all levels below the top level will be affected. That predator's normal prey items will proliferate, populations will not be held in check, food sources will be overexploited, and larger populations will be subject to the spread of disease. Catastrophic population declines are likely to occur.

We have seen these scenarios play out in the American West with the elimination of wolves, mountain lions, and similar predators. Marine biologists fear that the same potential exists for aquatic systems if sharks are eliminated. The sheer size of oceanic habitats would delay detection of these effects until the results were so dramatic that recovery could become impossible.

Commercial Products and Values

The greatest commercial demand for sharks is for their fins. When dried and prepared properly, they are used to make shark-fin soup, a concoction that is highly regarded, particularly in Asian markets. The fishery is worldwide and the quantity of sharks landed annually is staggering. Some estimates place the number of sharks taken annually at approximately 100 million. Other data suggest that number may be an underestimate and that the actual number may exceed 250 million. Actual numbers are difficult to obtain, because not all catches are legal and not all landings are reported to agencies that correlate worldwide fishing efforts.

Often the fins are taken and the remains of the shark dumped overboard while still alive; this practice is called finning, and the sharks soon die. Finning has been outlawed in most areas of the world. Discarding the bodies seems especially wasteful, since there is a

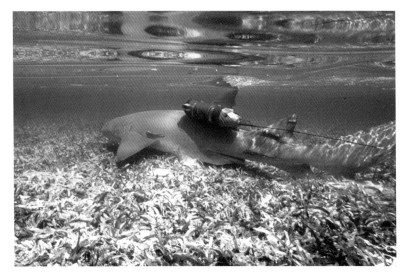

The CritterCam is an animal-borne video recorder that can record and store information showing exactly what the animals are experiencing. The device has a special electronic circuit that releases the camera from the animal when the recording time is complete. The device also has a radio, and its location can be determined from the radio signal. In spite of the size of the recorder, animals return to their normal behavior patterns even when wearing it. A nurse shark, in our studies of shark mating, is shown with a CritterCam. Also visible on its dorsal fin are Rototags and a transmitter, an array of shark jewelry that provides a rich bank of data from these animals. *Photograph copyright Harold L. Pratt. Used with permission. All rights reserved.*

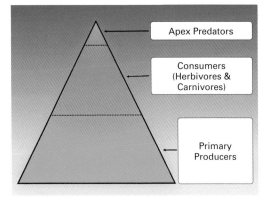

Food chains, food webs, and food pyramids describe the often complicated interrelationships between organisms that inhabit an ecosystem.

market for shark meat. When it is prepared properly, most diners find it to be tasty. A significant portion of the fish in "fish and chips" in European countries is derived from several shark species, and mako shark steaks have long been prized as a delicious seafood.

The fishery has become intense, and populations of sharks in some parts of the world are thought to be depleted so much that they might not recover unless stringent management initiatives are enacted and enforced. In areas where strong management practices have occurred, stocks are showing evidence of recovery, a sign of hope.

It's not just shark fins and meat that have some value. Ocean Leather Company of Newark, New Jersey, is often credited as the first commercial enterprise to develop a method of removing dermal denticles from shark skin to produce a high grade of leather, by a process that was patented in 1919 and 1920. The leather is quite attractive and relatively highly priced.

Ecotourism

Diving with sharks has become more popular in recent years, and tourist industries have grown worldwide to support this growing interest. Many resorts that cater to the recreational diver/photographer have gained immense popularity as the public's interest in encountering and studying these animals continues to develop.

The attitudes of government regulatory agencies toward sharks have begun to change in some positive ways. In certain developing countries especially, where shark fishing was once a serious industry, officials and entrepreneurs have come to realize that tourists attracted to large shark populations contribute substantial numbers of dollars. Cost analyses have shown that live sharks continue to generate a revenue stream for the long term, whereas harvesting is just a one-time income source.

Shark-fin soup is considered a delicacy in many cultures. Hunting and killing sharks to manufacture the soup explains why many shark populations have been overfished, some to the point where recovery may take generations.

Managed shark dives often use bait to attract sharks to provide opportunities for underwater photography. These dives are very popular and highly structured. While some bites and fatalities have occurred, they represent a small percentage of the individuals who experience the dives.

In addition to dive operations and sightseeing trips, resorts, restaurants, dive shops, and other commercial enterprises have sprung up in these locations and are supported by what have become thriving and lucrative dive and adventure destinations. However, not everyone is a fan of these types of dives. Some critics are afraid that sharks may associate food with all divers and, when no food is present, may turn on the divers instead. Baited dives, including cage diving, have been banned in some countries because of this fear and to avoid possibly altering the sharks' natural behaviors.

Attempts by Florida's agencies to regulate dives have not succeeded. The Bimini Tourism Advisory Board in the Bahamas has produced guidelines for tour operators to follow when diving with hammerheads, in hopes that the interactions with sharks, a large tourism industry, will continue with minimal impact on the shark population and maximum safety for divers. These guidelines are voluntary and do not carry the power of law.

Films

Sharks have a long history with Hollywood and have been featured players in films since the earliest days of filmmaking. Certainly *Jaws* is the best-known movie to feature a shark as its star. James Bond faced threats from sharks in the epic underwater scenes of *Thunderball*. Sharks have even been parodied in films, as when they rained from the skies in *Sharknado* and its many sequels. While the occasional appearance in movies might seem appropriate for a particular story line, in many films sharks have been featured simply to cater to the public's fear of them. Misrepresentation of sharks as a group by the media in quest of viewership does little to enhance the average person's understanding of the biology and positive roles of sharks in the oceanic environment. Finding the right balance between entertainment and education is a constant challenge for filmmakers.

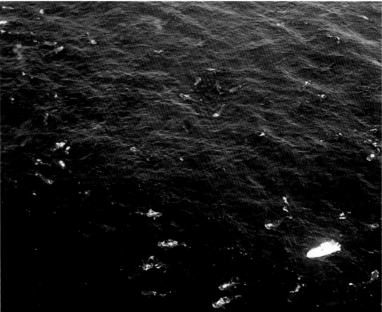

Whale shark aggregations offer worldwide opportunities to dive with the ocean's largest fish. Many governmental agencies have enacted protective legislation to protect a tourist industry that provides more income than harvesting the animals commercially. Researchers have begun to understand what drives these large migrations, and dive, snorkel, and whale-shark-watching excursions support a vast tourist industry. *De la Parra Venegas, R., R. Hueter, J. González Cano, J. Tyminski, J. Gregorio Remolina, M. Maslanka, A. Ormos, L. Weigt, B. Carlson, and A. Dove. 2011, An unprecedented aggregation of whale sharks,* Rhincodon typus, *in Mexican coastal waters of the Caribbean Sea,* PLoS One *6 (4): e18994, doi:10.1371/journal.pone.0018994.*

Conservation and Management

Fishing Regulations

Because of their vulnerability to overfishing, many shark species are protected and may not be taken by recreational boaters, fishers, or divers. Some of these species are protected from commercial fishing as well. Many other species are not under threat and may be taken, subject to size restrictions and bag limits. Because some of these species are actually endangered or threatened, special care must

Cage diving to view and photograph white sharks has become increasingly popular with a better understanding of where and when the animals can be dependably found.

Entire industries have developed to support the public's interest in diving with white sharks. Being in the water with a 16-foot white shark satisfies even the most steadfast adrenalin junkie's quest for adventure.

be taken to properly identify every animal, to be certain of its identity and protected status. Fines are large and jail terms are possible for mistakes in identification. The management of these species is under constant review. As a result, the regulations may change frequently, so fishers must consult the latest state and federal rules to stay current with regulatory changes.

In the Bahamas, the regulations are easy to understand: they ban all fishing for sharks. If sharks are caught by accident, they must be immediately released. In Florida, deciphering the rules and regulations is more complicated, because different regulations may apply to state waters, Atlantic waters, and the Gulf of Mexico. Trying to make sense of the various regulations and where they overlap can be a challenge. The federal government's advice "If you don't know, let it go" is probably a good rule of thumb to avoid issues

of identification if one is stopped by an enforcement officer. Regulations are constantly changing as new data emerge, so fishers are advised to make certain they are complying with the most recent regulations. Federal rules prohibit retention of the following eight Florida species, all of which are described in this book:

- Basking shark
- Caribbean reef shark
- Dusky shark
- Great white shark
- Sandbar shark
- Sandtiger shark
- Silky shark
- Whale shark

Florida laws prohibit retention of all of those species as well as the following five species:

Diving with sharks is only one way people can experience a shark encounter. Fishing for sharks is still popular, though it is highly regulated in most areas. Shark fishing is banned in the Bahamas, and many species are closely protected in Florida waters. For most of the species not prohibited, there are size limits and possession limits.

- Great hammerhead shark
- Lemon shark
- Scalloped hammerhead shark
- Smooth hammerhead shark
- Tiger shark

Though seldom sought as targeted species, many species of rays are also protected.

The continuing assessment of shark populations worldwide is a role of the International Union for the Conservation of Nature (IUCN). This organization brings together states, agencies of many governments, and nongovernmental agencies to gather worldwide data and statistics to assess and monitor population trends for shark species, as well as hundreds of other plant and animal species. More than 1,300 members from 185 countries work together to develop strategies and policies aimed at species protection, suggest laws where appropriate, and outline best-practice models. The agency then makes recommendations for management to ensure that healthy stocks will survive fishing pressures and that the populations of species under threat will remain or become stable.

Following their exhaustive study of a particular species, the IUCN places the species in one of several categories with recommendations for its management. Most species of sharks have been evaluated, and in the descriptions of species in part 2, the conservation status of each species is indicated. For comparison purposes, the categories, from least endangered to most endangered, are as follows. As can be seen from this list, an absolute determination for some species may be challenging and is not always clearly defined, especially for highly migratory aquatic animals. Furthermore, a particular species may be at risk in some areas but not necessarily throughout its entire range.

- Not evaluated—hasn't been studied against the various criteria
- Data deficient—not enough data, usually on distribution and abundance, to make a recommendation
- Least concern—usually widespread and abundant
- Near threatened—close to qualifying for a higher category or likely to qualify in the near future
- Vulnerable—faces a high risk of extinction in the wild
- Endangered—faces a *very* high risk of extinction in the wild
- Critically endangered—faces an extremely high risk of extinction in the wild
- Extinct in the wild—known only to survive under cultivation or in captivity
- Extinct—no reasonable doubt that the last individual has died

To Know Them Is to Understand Them Is to Value Them

To preserve and protect and sensibly harvest any living organism, it is imperative that we understand all aspects of its biology, including its role and contributions to the ecosystem it inhabits. Whether we're discussing coastal beach vegetation, particular forest trees, or fish in the ocean, education is a key element in wise and prudent use of any resource. This is no less true for sharks than for any other natural species. Because sharks have such a public relations problem and an image in need of repair, countless programs have been produced to familiarize the public with their value. Books, television shows, documentaries, and even weeks devoted to the study of sharks have, in many cases, begun a transition from fear to understanding and acceptance of the value of these supreme predators. The burgeoning shark ecotourism business is evidence supporting the public's desire to interact with these animals in controlled and nonthreatening encounters.

One example of an early education program was tested in Monroe County in the Florida Keys. Twelve middle-school girls were invited to compete in a pilot program that would give them the opportunity to participate in a basic shark research program. Based upon a proposal each one developed, participants were selected for a three-day program involving capture, tagging, and release of local shark species. The program was an attempt to provide the participants with an education based on firsthand experience with sharks and to challenge beliefs they had developed from the media portrayal of these animals.

The sessions began with an introduction to the fundamental aspects of shark biology, to familiarize the students with the animals they would encounter. In addition to basic biology, the students learned about the roles of sharks in the ecosystem as well as their commercial values. Open discussions were held to identify sustainable practices for fishing in light of the inability of sharks to rapidly replenish population numbers when fishing pressures become excessive.

After these sessions, the staff demonstrated how sharks that are captured in the study are measured, assessed for overall health, and tagged. Then each student was presented with one nurse shark, to direct the tagging process. Lengths and weights were taken, sex of the shark was determined, and external dorsal fin tags were applied in addition to internal microchip tags, very similar to what are commonly called pet identification tags. Tissue samples were taken for DNA analysis, and then the sharks were released.

The following day was devoted to fishing for sharks under permits from the Florida Fish and Wildlife Conservation Commission and the Florida Keys National Marine Sanctuary. Nineteen nurse sharks, four blacktips, one bull, and three bonnethead sharks were captured, tagged, and released during the 6 days. All animals were then released in the location where they were captured. One of the animals tagged by the students the day before was actually recaptured, demonstrating to the students how hardy and resilient these animals are and how their natural behaviors are unaffected by the tagging process.

Young women participants in a pilot Science, Technology, Engineering, and Mathematics program in the Florida Keys gain hands-on experience by catching, weighing, measuring and tagging sharks. The educational value of firsthand encounters helps to personalize the students' understanding of sharks and shark biology.

Tagging of an 8-foot bull shark by middle-school students brings the interaction with sharks to life.

During the initial introduction session, several students revealed their skepticism about working with sharks and their fear based upon fictional readings and television accounts that were more sensationalistic than truthful. In spite of their deep interest,

The shape of a hammerhead shark's head makes identification very easy. Distinguishing between the different hammerhead species, however, is a bit trickier.

they were concerned that shark research was inherently threatening. Comments and reflections of the students at the conclusion of the experience made it clear that those concerns had been set aside.

The production of educational media that is both entertaining and scientifically accurate is vital to an understanding of sharks. The sensationalism and blood and gore that lure viewers to the screen simply continues to play on fears of sharks and fails to educate and inform the public of the real value of these animals. It is truly a fine line for filmmakers to tread when they attempt to create a product that is both informative and entertaining, avoiding the boredom that many people associate with educational programming.

Perhaps workshops that target adults—though wholly unrealistic to create—would be the best way to transform attitudes and establish an entirely new understanding of this group of animals. Education, ideally incorporating personal experience, is essential to promote the understanding, protection, and preservation of these captivating animals.

Identifying Sharks

The identification of sharks is not always an easy task. Even experts frequently disagree. If you have a shark on a dock or even beside your boat, where intricate measurements can be taken, the task of identification is a bit easier. But for some that is not an option; beachgoers, surfers, divers, and even fishers sometimes struggle with distinguishing one species from another, especially when you are not close to the shark. With so many new laws and so many protected species, it becomes vital to correctly identify animals that are to be taken back to the docks, to avoid violating laws and incurring sometimes substantial fines.

Most Floridians and Bahamians who spend

time on the water can correctly identify many of the most common species. Hammerhead sharks, bonnethead sharks, and nurse sharks are among the easiest species to recognize. However, there are actually three species of hammerhead sharks, and distinguishing between them can be a challenge, especially if they are seen at a distance or by a diver. Having one hooked and next to the boat makes the task easier, but since they are a protected species, they are to be immediately released; thus, time cannot be wasted to take measurements or take enough pictures to make final identification a certainty.

Color is variable among animals and is not always a reliable way to identify different species, though in some instances it can be useful. For example, many juvenile sharks have fins with black edges and are often incorrectly identified as blacktip sharks.

The large white spots on whale sharks and their large size are sure clues to their identity. The spotting pattern is different between animals and can be used to identify individuals.

In other cases, color can be definitive. The white spots of a whale shark make it unmistakable.

If color isn't always reliable, what parts of a shark should a person observe to make a correct identification? For the species that this book covers, there are some useful guidelines that make it easier to narrow down identification of a shark to a few choices that can lead to a final decision. For the shark that only occasionally spends time in Florida and the Bahamas and may be rarely encountered, other identification guides will be necessary.

Knowing what to look for in advance is the best way to start developing the keen eye that will aid the decision-making process. Of course one has the option to get away from the shark and simply be satisfied that it was a shark. That might be sufficient for some people. Fishers, however, must be somewhat better informed.

Some characteristics are difficult to see unless the shark is very close. Often biologists rely on counting teeth in several rows or comparing the shape of dermal denticles, observations that most recreational fishers, divers, and boaters can't make. Similarly, comparative measurements of fin size and fin placement are not possible for the average diver or boater. "It was a big shark" might be enough to satisfy one's curiosity. For fishers or beachgoers who catch or happen upon a beached shark, more characteristics are evident.

The following few tips may help novices in the identification process. When these key characteristics are understood, a shark is easier to identify. If an absolute identification is necessary, a book specifically devoted to identifying the different species of sharks may be needed. Many volumes have been written that are dedicated to the finer details of identification. Those guides are useful references for readers who wish to expand their identification skills.

Not all of the traits are easy to spot at long

distances, but answering the following questions in a sequential fashion is a useful first step. For some species, such as the basking shark and the whale shark, the sheer size and unique body shape make identification very easy, and the following key points are not even necessary.

1. Are there distinct colors or patterns on the shark's body? Even though color is variable and not always a dependable identification tool, in some instances it may be useful. For example, stripes may indicate a tiger shark; a large shark with large white spots and an almost square snout may be a whale shark; a very small shark with a pointed snout and white spots may be an Atlantic sharpnose shark; and a shark with very deep black fin tips may be a blacktip shark or a spinner shark.

2. Is there a ridge on the shark's back between the first and second dorsal fins? If, so the shark is a "ridgeback" shark. Nine of the sharks commonly found in these waters are ridgebacks. The process of

Comparing the size of the dorsal fins of unknown shark species can be helpful in identification. The very nearly equal sizes of the first and second dorsal fins of the lemon shark are useful clues to its identity, since for most sharks, the first dorsal fin is larger than the second.

The shape of the tail is significant for identification purposes, especially whether the upper and lower lobe are the same size or if one lobe is longer than the other. The tails of the mako shark (*Isurus* sp.) and the great white shark (*Carcharodon carcharias*) are termed "lunate" because of their crescent or moon shape. In these species the upper and lower lobes of the tail are very nearly equal in size. As can be seen in the tail of this shortfin mako shark (*Isurus oxyrinchus*), there is also a ridge that extends out from the base of the tail. That part of the tail is called the caudal peduncle, and the ridge that extends outward is called a lateral keel. Both are diagnostic characteristics used to identify several shark species.

elimination can help to narrow down the choices within this group.

3. Are the first and second dorsal fins approximately the same size? Only three local species, nurse sharks, lemon sharks, and sandtiger sharks, have dorsal fins that are nearly the same size, and none of them is a ridgeback. Furthermore, these three species are relatively easy to identify and distinguish from one another.

4. Is the head shaped like a shovel or a hammer? Four sharks have this feature. The smaller bonnethead shark, with its rounded head, is very easy to recognize. The other three are the hammerhead sharks, and examining the front of the head can offer clues as to which hammerhead shark it is.

5. Is the snout blunt rather than pointed? Only three local sharks, lemon sharks, tiger sharks, and bull sharks, have obviously blunt heads. The tiger shark is a ridgeback (see number 1 above), and the lemon shark can be recognized by its equal-sized dorsal fins (see number 2).

6. Are the upper and lower lobes of the tail (the caudal fin) nearly the same size? Two local sharks, the mako shark and the great white shark, have this feature as well as very conical snouts.

The illustrations accompanying the species accounts can also aid in identification.

The following accounts will help readers identify sharks that may be seen or captured, but they do not offer comprehensive keys to identification. Shark identification may depend on measurements or examination of tissues that can only be obtained from dead animals. For some species that are protected by law, more detailed information is provided so that errors can be avoided for sharks that are legally taken for food. Readers are encouraged to consult the information provided by local, state, and federal law enforcement agencies and the publications they provide to aid in definitive identification.

PART TWO

Selected Shark Species

The accounts of individual species in this part of the book are intended to provide some basic and intriguing facts about these animals to readers who are boaters, fishers, swimmers, surfers, beachgoers, and divers, as well as others who are either basically familiar with sharks or wish to learn more about them. They are not intended for the professional biologist or fisheries manager. There is a separate body of advanced literature to satisfy people who have more than basic curiosity.

For each species, its frequency of occurrence is indicated. Not all of these species are easy to find in all Florida and Bahamian waters. The term "common" is used for sharks that are found very often. Even so, some species are more common in some regions than others. Some are found near shore; some are open-ocean species; some are more tropical; and some are most often encountered in the northern Gulf of Mexico or the northeastern Florida coast and are less common in the Florida Keys or the Bahamas. Other sharks are "less common" because of their restricted distribution, and some are considered rare because they are not often found in these waters or they are encountered only on their way to feeding, breeding, or nursery grounds. Nevertheless, it is the intent of these accounts to help people who might encounter sharks in Florida and the Bahamas to identify the animals they do

encounter and to answer the questions that most often arise when a shark makes an appearance.

It should be noted, with regard to the basic statistics, that there is not 100% agreement among scientists or fishers regarding the distribution, sizes, and similar values for the sharks. Thus each account provides a range of measurements and statistics to reflect the most commonly cited values.

Each species account discusses human interactions with these species, such as diving and fishing encounters, including managed shark dives or shark tournaments as well as the categories too commonly referred to as attacks. The attack data were kindly provided by George Burgess and Lindsay French at the Florida Museum of Natural History at the University of Florida, the location of the International Shark Attack File. The information in this large database recorded 743 Florida shark attacks from 1882 to the present and 27 attacks in the Bahamas. In only 15 of the Bahamian attacks was the identity of the attacking species confirmed. All of these attacks were unprovoked. In only 98 of the Florida attacks was the attacking species positively identified. If a species has been identified from this file, that information is included in the appropriate species account.

The conservation and management status of each species is also reviewed. The source of these data is the International Union for the Conservation of Nature (IUCN). That organization brings together representatives from countries worldwide, from governments and from nongovernmental agencies, that work collaboratively to evaluate species with the goal of making recommendations to ensure their survival. From its own description of its charge, the IUCN

> has been assessing the conservation status of species, subspecies, varieties, and even selected subpopulations on a global scale for the past 50 years in order to highlight taxa threatened with extinction, and thereby promote their conservation. . . . The IUCN Red List of Threatened Species™ provides taxonomic, conservation status and distribution information on plants, fungi and animals that have been globally evaluated using the IUCN Red List Categories and Criteria. This system is designed to determine the relative risk of extinction, and the main purpose of the IUCN Red List is to catalogue and highlight those plants and animals that are facing a higher risk of global extinction (i.e. those listed as Critically Endangered, Endangered and Vulnerable). The IUCN Red List also includes information on plants, fungi and animals that are categorized as Extinct or Extinct in the Wild; on taxa that cannot be evaluated because of insufficient information (i.e., are Data Deficient); and on plants, fungi and animals that are either close to meeting the threatened thresholds or that would be threatened were it not for an ongoing taxon-specific conservation programme (i.e., are Near Threatened). (IUCN 2016)

The Shark Specialist Group of this organization consists of biologists and fishery managers from around the world who evaluate all known science, historical, and contemporary commercial and recreational fishing data and formulate guidelines for every shark species that may be facing some risk to its population stability. The recommendations are forwarded to governments and international organizations that formulate policy and establish protective measurements, many with protection of international law, to guide all nations in their resource management policies.

Atlantic Sharpnose Shark

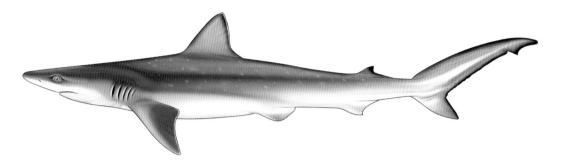

Common Names: Atlantic sharpnose, sharpnose

Scientific Name: *Rhizoprionodon terraenovae*

Identifying Features: The Atlantic sharpnose shark is a small, slender shark. It seldom exceeds 40 inches (100–110 cm) in length. It is generally brownish to gray in color and often has a metallic sheen with white spots on the upper surface that may fade to a whitish color on the sides and underside. Mature animals without scars are quite attractive. The edge margins of the dorsal fins and tail may be black. The shark's slender shape and very pointed snout make it comparatively easy to identify.

Frequency: common

My first experience with Atlantic sharpnose sharks came on a fishing trip to the backcountry islands, north of the Florida Keys. The trip was designed to catch and tag small nurse sharks to see where they moved in the middle and lower Keys. The first task on these sampling trips is to catch and prepare baits for nurse sharks. Pinfish, small jacks, and small snappers make ideal baits, along with fresh mullet. Usually live shrimps or small crabs work well to entice these baitfish to the boat, especially if a block of frozen chum is in the water to lure them to the back of the boat. Short casts with live shrimps

were effective, but occasionally a small bonnethead shark would interrupt the tedium of catching pinfish and would provide a pleasant fight on light tackle.

One particular cast was followed immediately by a ferocious strike and an initial run that made me think I had hooked a bonefish or a small tarpon. The run was short, however, and I began to gain back the line stripped from the reel. As the animal came closer to the boat, I could see that it was a shark. I first thought it was a small lemon shark because of its sleek appearance. As it came closer, however, I saw that its snout was very pointed and that small white spots dressed its dorsal surface. When I netted the shark prior to its release, I could see that it was an Atlantic sharpnose shark. I was struck by its intricate coloration, almost as though it was airbrushed. The fight did not take long, though its initial effort was worthwhile. The animal was released and easily swam back to its seagrass bed.

The Atlantic sharpnose shark is very plentiful in Florida waters, almost to the point of being considered a pest by nearshore and surf fishers who prefer to catch fish other than sharks. It is among the most abundant sharks in the Gulf of Mexico but is generally not encountered as frequently in the Bahamas.

Commercial fishers who seek other species regard the Atlantic sharpnose shark with the same disdain as fishers in the North Atlantic

The white spots that grace the flanks of the Atlantic sharpnose shark are one of the distinctive markings that help to identify this shark.

The sharply pointed snout of the Atlantic sharpnose shark makes it easy to understand how it earned its common name.

regard the spiny dogfish shark (*Squalus acanthias*). Both species tend to foul nets and require a great deal of extra effort to free them. The spines of the spiny dogfish, however, are much more of a problem to commercial fishers, since they easily tangle nets and their spines can cause hand and finger injuries. The Atlantic sharpnose shark does not have spines.

Range, Distribution, and Habitat Preference

Atlantic sharpnose sharks are abundant on the Florida east coast and in the Gulf of Mexico. They are frequently found in schools in bays, estuaries, and other shallow waters, even penetrating into brackish waters. Fishers who enjoy eating sharks consider the Atlantic sharpnose shark a tasty fish. The sharks prefer shallower waters, but specimens have been taken from depths of several hundred feet. Tracking the animals with special transmitters shows that they move from estuary to estuary through deeper waters. Other studies show that they spend the warmer months near shore and then move to deeper water in the winter months. Males tend to move away from shore when water temperatures become too warm.

Size, Age, Growth, and Reproduction

Atlantic sharpnose sharks are small, seldom reaching more than 40 inches (101 cm). Their young are born at a length of 11 to 12 inches (30 cm). They grow rapidly in their first year, doubling their size to 2 feet (60 cm). Their growth rate slows thereafter, and during their third year they may grow only around 6 inches (15 cm). This is the approximate size and time at which they reach sexual maturity. Both males and females reach sexual maturity at a length of around 24 to 31 inches (60–80 cm). They are thought to reach their maximum length of 40 inches (101 cm) shortly thereafter and may live 8 to 10 years.

Young sharpnose sharks are born live.

A placenta connects developing embryos to the mother during the development of the pups. Litters generally consist of four to seven offspring, and gestation may be as long as 1 year. Females are thought to mate very soon after giving birth and reproduce annually.

Food and Feeding

The Atlantic sharpnose shark prefers to feed on small fish and crustaceans such as crabs or shrimps. Food preferences may vary depending on where the animal is located, so the sharks show some flexibility in their diet. Examinations of their stomach contents have found mostly fish remains, though some studies in different locations have found mostly crabs and shrimps.

Behavior and Interactions with Humans

Because the Atlantic sharpnose shark prefers shallow, inshore, coastal waters, these sharks may frequently be seen by beachgoers or surfers. Incidental contact may occur, and some small bites have been reported that are generally not serious. Compared to other larger and more aggressive species, these small, shy sharks have few interactions with humans.

Conservation and Management Status

Atlantic sharpnose sharks are considered a small coastal species for management purposes. They are popular in both recreational and commercial fisheries and are often caught in trawls as bycatch. Nevertheless, their ability to reproduce rapidly protects them well from the risk of overfishing in Florida, provided that monitoring and quotas continue to protect the stability of the population. There is no minimum size for recreational fishers.

The IUCN rates their status as of "least concern." In other parts of their range, including Mexico, they are more heavily fished and may possibly require more stringent protection over time, depending upon changes in fishing pressures, particularly from artisanal fisheries.

Basking Shark

Common Names: basking shark, basker, sunfish

Scientific Name: *Cetorhinus maximus*

Identifying Features: Basking sharks are very large animals and generally grayish-brown in color. They can be easily distinguished from whale sharks because they lack the white spots that characterize whale sharks. They also possess a bizarre head and a conically shaped snout that gives them an almost comical appearance when feeding. The gill slits are also very large, nearly enveloping the entire circumference of the head.

Frequency: rare

Imagine a miniature submarine with a double screen door for a bow, the ability to float at the surface, and a conning tower that resembles a huge oval fin. That design, with some minor biological adjustments, is a pretty accurate description of a basking shark moving through the water and feeding. The basking shark ranks as the second-largest fish species. Only whale sharks are larger. Both species share the same filter-feeding strategy, and their huge size and gaping mouths make them easy to identify.

Basking sharks are frequently sighted at the surface swimming with their large mouths agape as they strain plankton from the surface waters. Their habit of swimming on the surface may be responsible for their common name, since it appears that they

Basking sharks feed on plankton in the surface layers of the ocean.

are basking in the surface sunshine. It is not uncommon to encounter large schools of basking sharks working through dense aggregations of floating plankton.

Range, Distribution, and Habitat Preference

Basking sharks prefer cooler, temperate waters but are occasionally sighted in Florida and the Bahamas. While there are records of sightings in the Gulf of Mexico, the sharks are more likely to be spotted on the Florida east coast and in the Bahamas. This may possibly coincide with long-distance migrations taking them from the cooler waters of the northeastern US coast through the Caribbean to South America. Some reports suggest they may even move as far south as Brazil and Argentina. Their appearance in Florida waters may be only as transients as they follow migration paths.

Satellite telemetry is just beginning to reveal the details of these large-scale movements. Early evidence from such studies has shown that basking sharks are capable of long-distance migrations including transoceanic passages. Several animals were actually tracked moving from the British Isles across the Atlantic to Newfoundland, Canada, a distance of 5,993 miles (9,589 km). During this crossing, the animals reached depths of 4,145 feet (1,264 m), depths previously unheard of for these sharks.

Size, Age, Growth, and Reproduction

As the second-largest fish in the sea, basking sharks are thought to exceed 40 feet (12 m) in length, although measured specimens are more often in the range of 23 to 30 feet (7–9 m). As with many of the other large marine animals, exact length measurements are difficult to obtain and estimates are commonplace. More exact measurements are generally made from captured animals, most commonly from commercial fisheries when dead animals are often measured before preparation for the market. Whether these

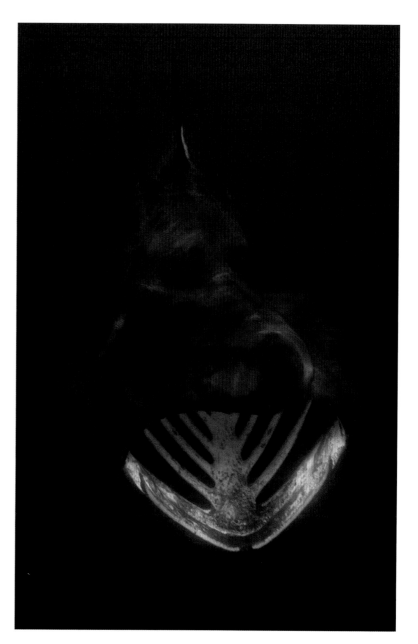

The large size of the basking shark's mouth would be more fearsome if the sharks were not plankton feeders.

measurements represent average sizes is not certain.

Very little is known about age and growth in basking sharks. One study was able to determine that one female grew a total of slightly more than 7 feet (2.4 m) in 3.1 years. This would equate to an annual growth estimate of nearly 30 inches (74 cm) per year. Small basking sharks of 5 to 6 feet (150–180 cm) have also been reported.

The large gill slits of basking sharks nearly encircle the head and serve as an identifying characteristic to help to distinguish basking sharks from the other large shark species.

In addition, little is known of the mating and reproduction of basking sharks. The size at which they reach sexual maturity has been estimated to be between 13 and 16 feet (4–5 m) in males and more than 26 feet (8 m) in females. Pregnant females are not often encountered in targeted fisheries, so little is known of gestation and the size of the young at birth. Some biologists have speculated that the gestation period could be as long as 3 years, an exceedingly long period to be pregnant, for any species. There does not seem to be universal acceptance of this hypothesis. Embryos obtain their nutrition from a large yolk sac during development and may also consume other eggs that continue to be produced by the mother. There is no placenta connecting the developing embryos to the mother.

Food and Feeding

Basking sharks are most likely to be encountered swimming slowly at the surface with their large mouths agape. As they swim, water flows through the mouth and the gills. The sharks rely on this passive flow of water

to provide food as they filter huge volumes of plankton-rich water during their feeding forays. Unlike whale sharks, basking sharks do not suction feed. Their oversize gill slits may serve the additional advantage of permitting increased volumes of water to pass through the mouth and across the gill rakers, thereby enabling them to filter larger quantities of planktonic prey items from the water column.

Known to migrate over long distances, basking sharks may actually move in response to changes in plankton densities. Copepods make up a major portion of their diet, and how basking sharks are able to sense or predict such changes in copepod population densities remains a mystery.

Interactions with Humans

Encounters between divers or swimmers and basking sharks are rare. Underwater photographers have been able to obtain stunning images of the animals as they swim with their large mouths agape, and the sharks do not seem to be bothered by these experiences. In spite of this indifference to humans, their large size should be reason for caution. An unexpected bumping from one of these large sharks could potentially injure a diver or swimmer, and a collision with a small boat might result in unintended consequences.

Conservation and Management Status

Once basking sharks were sought for their gigantic livers, from which useful oils were extracted, but fishing for these sharks has diminished in many locations as a result of international efforts to protect dwindling stocks. Some limited fisheries still exist for the oil, flesh, and skin of the basking shark, all of which are utilized in the shark marketplace. Many of the animals that are caught today are considered as bycatch from bottom trawl fishing. The reduction in their numbers, however, makes the directed fishery impractical. Efforts to establish more stringent

Basking sharks have been called the gentle giants of the sea since they move and feed with what seems to be little effort.

worldwide protections have not been successful, though the United Nations is actively attempting to establish better monitoring and management initiatives for several species of sharks, including basking sharks.

The IUCN currently lists basking sharks as "vulnerable," which means that fishing efforts must be closely monitored to prevent the numbers from decreasing to such low levels that recovery becomes impossible.

Blacknose Shark

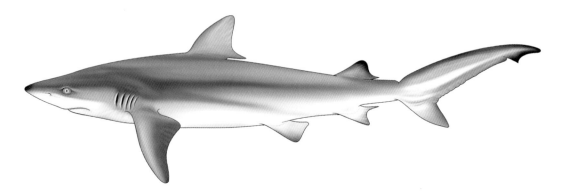

Common Name: blacknose
Scientific Name: *Carcharhinus acronotus*
Identifying Features: The blacknose shark is gray-green or yellowish gray on the upper surface. The second dorsal fin and the upper tip of the tail may also have blackened tips. There is no ridge between the dorsal fins. Blacknose sharks seldom grow very large. Their most distinctive characteristic is a dark shading on the tip of the snout that gives them the appearance of a black nose. This coloration may fade as the animal ages and is not always a reliable criterion for identification. The tip of the tail and the second dorsal fin may also show the deeper coloration and may help with identification. Because blacknose sharks seldom reach a large size, any small sharks with this color pattern may be blacknose sharks.
Frequency: common

The blacknose shark may best be described as a small shark with an attitude problem. Though they are relatively shy and are seldom seen by divers, feeding operations to attract sharks for underwater photography occasionally attract them. Divers who have been able to photograph them have reported that the animals will often give a threat display by arching their back and pointing their fins dramatically. In spite of this behavior—bravado for such a small shark—no attacks on humans have been reported.

Range, Distribution, and Habitat Preference

The blacknose shark is widely distributed throughout the southwestern Atlantic, the Caribbean Sea, and as far south as Brazil. It prefers shallow tropical and warm temperate water and is most frequently encountered near the shore. The blacknose shark is also a common species in the Gulf of Mexico.

The bronzed coloration of the blacknose shark stands out as it cruises the waters above a Bahamian reef.

The blacknose shark is a very common small shark in the coastal waters of Florida and the Bahamas. Blacknose sharks are more likely to be caught by fishers working very near the shore, where they are considered to be among the most common small sharks, particularly along the Florida east coast. They have a preference for sandy bottoms or areas with shell and coral rubble.

Age, Growth, and Reproduction

The maximum size of blacknose sharks is reported to be 55 to 59 inches (140–150 cm), though one exceptionally large animal was measured at almost 6.5 feet (200 cm). The average size is much shorter. Their life expectancy is thought to be 16 to 19 years. In their first 4 to 5 years, they grow 20 to 25 inches (50–65 cm) before they become sexually mature.

There are some interesting differences in age, growth, and maturity depending on where the animals live. Populations in the South Atlantic reach sexual maturity at an earlier age and live longer than animals from the Gulf of Mexico. No explanation has been given for these differences.

Blacknose sharks are thought to reach

Blacknose sharks are common visitors to the shallow sandy flats, where they are often found in small schools in the crystal-clear flats.

The smudge that is found on the snout of the blacknose shark fades as the animals age.

Blacknose sharks are common residents of inshore reef systems.

A curious blacknose shark basks in the sunlit shallow waters of south Florida and the Bahamas.

sharks. Newborn blacknose sharks are 11 to 13 inches (31–36 cm) long, though some studies have shown that newborns may be somewhat larger, perhaps up to 17 inches (43 cm) long. Females reproduce at 2-year or longer intervals.

Food and Feeding

The blacknose shark is a speedy fish and is highly regarded as a strong fighting fish on light tackle. It is generally a fish-eating shark. Small inshore fish such as pinfish, porgies, anchovies, and pufferfish have been found in blacknose sharks' stomachs. Blacknose sharks have also been found in the stomachs of other sharks, suggesting that they have natural enemies of their own.

Behavior

Blacknose sharks have been known to form large schools. There is some segregation by sex, so the schools may be more dedicated to feeding than to mating. No attacks on humans have been recorded. However, threat displays, even from small sharks, should not be ignored.

Conservation and Management Status

Blacknose sharks have some value to the commercial industry and are fished throughout their distribution. They are also taken in significant numbers as bycatch by deep trawls, by shrimp trawls, and by coastal gillnets. However, fishing efforts are not thought to be heavily impacting their populations, and they are not considered at serious risk. Their populations are being continuously monitored to observe changes that might eventually warrant extra consideration. In Florida there is no size limit, and only daily bag limits apply to recreational fishers. The IUCN lists the blacknose as "near threatened."

sexual maturity at lengths of 35 to 39 inches (89–96 cm), when they are about 4 to 6 years old. In some parts of their distribution, they have been reported to be somewhat larger before they become sexually mature. Females are pregnant for 10 to 11 months and give birth to one to five pups. There is a placental attachment to the mother from the embryos during the development of the embryonic

Blacktip Shark

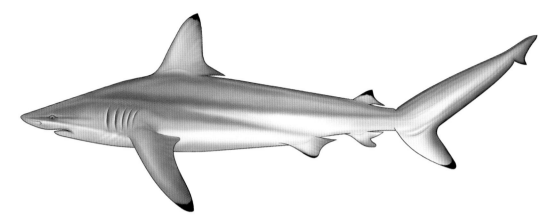

Common Name: blacktip
Scientific Name: *Carcharhinus limbatus*
Identifying Features: The blackened fin tips are the most obvious distinguishing characteristic of blacktip sharks. However, the black on the fins may fade with age, making identification difficult. Young blacktip sharks are easily confused with many other small coastal sharks, especially spinner sharks. Their anal fin is light or white, and this helps to distinguish them from spinner sharks, which are very similar in shape and coloration but have a black-tipped anal fin. The body color of a blacktip shark is brownish gray, tending to a metallic bronze. The snout is pointed. The pectoral fins are pointed. There is no ridge on the back between the dorsal fins, and a white band extends alongside the flank.
Frequency: very common

Blacktip sharks are among the most common sharks in the shallow inshore waters of Florida and the Bahamas. The black coloration on the tips of the fins is an easy way to identify them. However, so many juvenile sharks carry the same markings that juveniles of many species are incorrectly identified as blacktip sharks. This could be a problem for fishers who land small lemon sharks, which are protected in Florida and often have blackened fin tips. The almost equal size of the lemon shark's dorsal fins is an easy way to identify them, since a blacktip shark's second dorsal fin is very small compared to its first dorsal. Nevertheless, to avoid possessing protected species, it is important for fishers who target these small sharks for food to learn to tell the difference between similarly marked juvenile and small sharks.

Blacktip sharks are highly prized by game fishers and commercial fishers alike. Their

Blacktip sharks are among the most common sharks in Florida and the Bahamas. The blackened tips of the fins make them easily recognizable. They are often confused with spinner sharks, though the white anal fin of the blacktip shark is different from the black-tipped anal fin of the spinner shark.

meat is tasty and considered commercially valuable. Only the sandbar shark is more highly sought. Blacktip sharks are also prized as gamefish, often jumping clear of the water and putting up a strong fight on light tackle.

Range, Distribution, and Habitat Preference

Blacktip sharks are found worldwide. They prefer temperate and warm tropical waters and are extremely common throughout Florida and the Bahamas. They are commonly discovered near shore and in shallow bays and estuaries, preferring these shallow habitats to deeper water. They are frequently and easily caught by fishers in such areas. Blacktip sharks may be the most abundant of the larger coastal species that are found in the Gulf of Mexico.

Blacktip sharks are often found in large schools and gather for feeding and migratory movements.

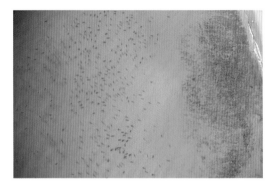

The migratory patterns of blacktip sharks have been studied as they travel close to the Florida coasts. Heavy concentrations of the sharks are present during the spring months. They have been studied using aerial photography along the southeastern coast. *Photograph courtesy Mark Mohlmann. Used with permission.*

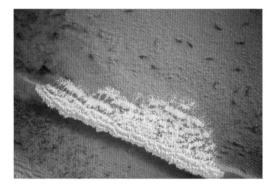

Blacktip shark migrations frequently bring large schools of sharks near the shoreline, just outside the surf zone, as they migrate along the Florida coastline. *Photograph courtesy Mark Mohlmann. Used with permission.*

Schooling blacktip sharks provide ample opportunities for underwater photographers.

Size, Age, Growth, and Reproduction

Blacktip sharks are not extremely large sharks, seldom reaching lengths over 6 feet (2 m). However, there have been some reports of animals measuring nearly 8 feet (2.45 m). Males reach sexual maturity when they are about 5 feet (145–150 cm) long, and females at a few inches longer. Males are 4 to 5 years old when they mature, and females are 7 to 8 years old. Their life expectancy has been estimated to be around 12 years.

Extensive studies of Gulf of Mexico populations of blacktip sharks indicate that they grow very rapidly in their first year, perhaps even doubling their length. Thereafter the rate drops until just after they mature, when they attain lengths close to their maximum size. As full-grown adults, their growth almost ceases.

Blacktip sharks are live-bearing sharks. Their gestation period is approximately 1 year, and they are capable of reproducing every other year. Litter size is usually 4 to 6 pups, but larger litters of up to 10 have been reported. The pups range in size from 20 to 23 inches (50–60 cm).

Blacktip sharks are frequently observed in classic feeding frenzies at fish cleaning stations or other places where large amounts of bait or fish remains are in the water.

Nursery grounds are generally in shallow inshore bays, where the pups may reside for most of their first year. Studies have shown that young blacktip sharks may aggregate in large schools during the daytime and disperse in the evening. Whether this schooling behavior offers them protection from predators or whether it enhances feeding opportunities is unknown. In spite of these protective behaviors, more than half of the newborns may

The perfect shape and coloration of the black-tip shark make it an ideal subject for under-water artists.

not survive longer than 15 weeks—a very high rate of mortality.

Blacktip sharks have also been shown to exhibit parthenogenesis, the production of offspring by a female without sexual contact with a male. The bonnethead shark is another of the very few species that have been shown to be capable of this type of reproduction.

Food and Feeding

In Florida and the Bahamas, blacktip sharks prefer fish, including fish found in shallow inshore bays and estuaries, such as mojarras, small snappers, and mullet. Common reef fish such as grunts, porgies, triggerfish, jacks, and groupers have also been seen in stomach-content analyses. Blacktip sharks will also prey on small sharks and rays. In other parts of their distribution they consume menhaden, anchovies, sardines, and other schooling fish.

Behavior and Interactions with Humans

Schooling blacktip sharks are frequently observed near shore, just outside the surf zone, when they undergo their migrations. News reports and helicopter videos of these massive schools do not delight chambers of commerce, as schools in the thousands are shown adjacent to popular beaches with swimmers seemingly just feet away from the sharks. Commercial fishers will often take advantage of these large numbers to increase their catches of this commercially valuable shark. Recreational surf fishers also delight in the large numbers of animals, though possession limits imposed in Florida restrict the total number of sharks that may be taken. For recreational fishers there is no size limit for blacktip sharks.

Blacktip sharks are blamed for many incidents with humans. Because of the sharks' preference for shallow waters, and especially surf zones where the water is turbid, smells are dispersed, the waters are noisy, and collisions with swimmers and surfers may be common. In those collisions, it is believed that the sharks slash out when incidental contact occurs, perhaps inflicting a nasty wound, and then quickly depart. The International Shark Attack File lists 14 confirmed unprovoked blacktip shark attacks in Florida, all of which were nonfatal. Fifteen attacks have been attributed to the spinner shark and several unconfirmed attacks were thought to be from blacktip sharks, though the absolute identity was not confirmed. Only the bull shark is credited with more attacks (19) than blacktip sharks. No attacks attributed to blacktip sharks have been reported from the Bahamas.

Conservation and Management Status

Because the blacktip shark's flesh is so highly prized, the recreational and commercial fisheries for blacktip sharks are closely monitored. This shark is considered to rank second in terms of its value to the fishery. Only sandbar sharks are considered more valuable. However, Australia and the United States are the only countries to enact management plans that include blacktip sharks. Although the first-year mortality of the blacktip shark is relatively high, its rapid growth to maturity and favorable reproduction results seem to offer some level of protection to the populations. As such, it is deemed "near threatened" by the IUCN. A species is near threatened "when it has been evaluated against the criteria but does not qualify for Critically Endangered, Endangered or Vulnerable now, but is close to qualifying for or is likely to qualify for a threatened category in the near future" (IUCN 2012, 15).

Bonnethead Shark

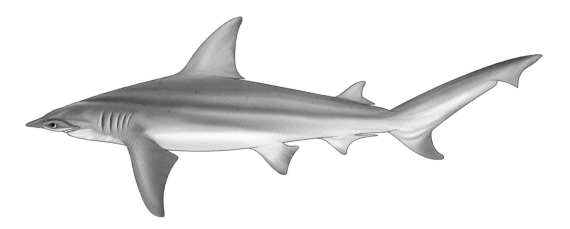

Common Names: bonnethead, bonnet shark, shovelnose, shovel head shark
Scientific Name: *Sphyrna tiburo*
Identifying Features: A bonnethead shark has a small, rounded head resembling a shovel. The eyes are located at the edge of the head. The sharks are gray in color, often with small black spots. Bonnethead sharks are true hammerhead sharks and are easy to distinguish from other sharks. The rounded head also makes them look different from other hammerhead sharks, whose heads are not rounded but flat across the front of the snout. They are not large sharks and seldom grow to more than 60 inches (150–160 cm).
Frequency: very common

The smallest of the hammerhead sharks, bonnethead sharks are among the most common sharks found in the shallow inshore waters of Florida and the Bahamas. They are accomplished hunters and voracious feeders. They will readily attack a bait and provide a short but spectacular fight on light tackle. Like other hammerhead sharks, they fight so hard that they often die from the stress unless they are released immediately. They often have to be resuscitated by holding them behind the head and moving them through the water to keep a current across the gills until they recuperate and can be released.

Bonnethead sharks are often seen in display aquariums and seem to do well in captivity, if care is taken during capture and transport. While they are common in nearshore waters, they are shy around divers and are difficult subjects to photograph underwater.

The shovel shape of the bonnethead shark makes it easy to recognize and easy to distinguish from other hammerhead sharks.

Bonnethead sharks are not very cooperative subjects for photography. They are shy and fast and are easily able to defeat most attempts to capture them free-swimming in open water.

Range, Distribution, and Habitat Preference

Bonnethead sharks are found from Rhode Island to Brazil, though they prefer warmer waters. They are extremely common in Florida and the Bahamas and are generally encountered inshore of continental and insular shelves, where waters are shallower and warmer. They are commonly found over sand, mud, seagrass beds, and marsh and mangrove habitats. They are also common on and around coral reefs. Sport fishers in shallow water who are fishing for bonefish, permit, and tarpon will frequently catch bonnethead sharks, especially if live shrimps or small baitfish such as pilchards or pinfish are being used as bait.

Size, Age, Growth, and Reproduction

Bonnethead sharks are the smallest of the hammerhead sharks. Their rounded heads make them easy to distinguish, even from juvenile hammerhead sharks of the other species that are the same size. Their average length is 3 to 4 feet (90–120 cm), though smaller animals are very common. The larg-est reported size is around 5 feet (150 cm). They are often encountered in schools that may number in the hundreds. Whether these huge schools are feeding aggregations, mating aggregations, or groups preparing to migrate is unknown.

Bonnethead sharks may grow as much as 10 to 14 inches in their first year. They reach sexual maturity at a length of 22 inches (55–70 cm) and an age of 2 years for males and, for females, 31 inches (80–84 cm) and an age of 2 to 3 years. Males may live as long as 16 years, and females may survive for up to 17.9 years.

Bonnethead sharks are viviparous sharks with a placental attachment. The gestation period is thought to be 4 to 5 months, though they have the peculiar ability to store sperm for some time before fertilization occurs. Litters have been shown to have multiple fathers in some instances. Females are thought to reproduce annually.

Young bonnethead sharks are born at lengths between 9 and 13 inches (24–34 cm) by some estimates, and 20 to 29 inches (52–75 cm) in other studies. Animals captured in the Keys that have spontaneously aborted near-term litters are much smaller, with average sizes more in the 10–12-inch (25–30 cm) size range. Newborn pups birthed in captivity are preyed upon by great white herons, who seem to find the baby sharks particularly tasty. As bonnethead sharks grow, they become prey items for other sharks as well as large groupers.

Bonnethead sharks share one very unusual characteristic for sharks. Several of them have been shown to give birth by the process of parthenogenesis (without involvement of a male).

Food and Feeding

Gut content studies list shrimps, crabs, small mollusks, including octopuses, and small fish as principal food items. In some regions bonnethead sharks have a strong preference for blue crabs. They are extremely fast swimmers and are efficient at tracking prey. Small fish would have difficulty in escaping a hunting bonnethead shark.

Behavior and Interactions with Humans

Large schools of bonnethead sharks are frequently seen in the shallow waters of the Florida Keys, for reasons unknown. Animals in these schools seldom show mating scars. The migration of bonnethead sharks has been well studied; long-distance movements may not occur. Long-term tagging has shown that they will faithfully return to areas where they are known to feed, even after intervals as long as 9 years. In one study, some animals that had been initially tagged at the same time in the same location were later all recaptured in the same location after intervals as long as 3.6 years. This study is used as evidence for what is known as site fidelity (preference for and faithful return to a preferred area) and group cohesion (staying together as a group).

Bonnethead sharks are shy, and opportunities for interaction with humans are rare. Snorkelers in shallow bay or mangrove areas may occasionally sight one, but the encounter is generally very short. The more common interaction occurs during fishing, when they are taken as bycatch.

Conservation and Management Status

Some bonnethead sharks are taken by recreational fishers for food, and they are taken in large numbers by commercial fishers, though the total catch, including bycatch in shrimp trawls, is apparently not considered sufficiently high to warrant protective measures. The IUCN lists the species as of "least concern" because of its ability to rapidly replenish its numbers; it has a life history that includes early maturity, high litter sizes, and short gestation times. Bonnethead sharks are not regulated in Florida and have no size limit.

Bull Shark

Common Names: bull shark, ground shark, cub shark, Lake Nicaragua shark, Zambezi River shark

Scientific Name: *Carcharhinus leucas*

Identification: Bull sharks are large, very stocky, deep-bodied sharks often described as stout or robust. Their snouts are very short and rounded. They are very similar in overall structure to several other shark species, including Caribbean reef sharks and dusky sharks and are often hard to tell apart from these species. Caribbean reef sharks and dusky sharks have a ridge on their dorsal surface between the first and second dorsal fins. Bull sharks do not have this ridge. There are also some differences in the fin size, shape, and placement. Identification is easier with animals that have been captured, especially when compared alongside species that they closely resemble.

Frequency: common

Imagine a day on a raft or small boat in the Mississippi River trying to duplicate the adventures of Tom Sawyer and Huck Finn. After you cast a small bait into the water, hoping for a dinner-sized catfish, your line, cane pole, and anything else in the water simply disappears. Your freshwater expedition may have turned into more of an adventure than you expected, especially if you were able to see that your trophy catfish was in fact a small bull shark.

There are many stories and official records of bull sharks being caught hundreds of miles up the Mississippi River. One animal was verified by the Illinois Department of Conservation to have been caught in Alton, Illinois, in 1937, 1,750 miles up the Mississippi. Reports of bull sharks from several thousand miles up the Amazon River also exist. It should come as no surprise, then, that bull sharks have been routinely captured and tagged in the rivers and creeks of the Florida Everglades, many miles up the St. Johns River in Jackson-

The blunt snout and long, pointed pectoral fins are visible in this bull shark profile.

ville, and in almost all lagoons and bays along all the Florida coasts. While other species of sharks can and often do make periodic runs into freshwater, the bull shark is the best known and most commonly encountered shark in freshwaters of the United States.

There is some disagreement about what shark species is the most dangerous to humans. The great white shark gets the most press and has a terrible reputation. But many people, including many shark biologists, believe the bull shark may be even nastier. Its nearly worldwide distribution and its prefer-

ence for nearshore waters may bring it into contact with more people than the number who encounter great white sharks. The International Shark Attack File lists many attacks attributed to bull sharks. Since they inhabit many desolate areas and are common in Third World countries, reports of attacks in those remote areas may not even be made, so the actual number of bull shark attacks may be much larger than official records indicate. There is no question that bull sharks are dangerous sharks.

Range, Distribution, and Habitat Preference

Aside from their occurrence in Florida and the Bahamas, bull sharks are found worldwide. The shark's many common names reflect its wide distribution, from Lake Nicaragua in Central America to the Zambezi River in Africa and throughout the Pacific and Indian Oceans.

Bull sharks prefer shallower waters inside the continental shelf in tropical and subtropical regions. They are frequently found near shore and commonly penetrate into freshwater rivers. Their preference for estuarine and fresh waters may be especially important

Photographers who find themselves in the water with bull sharks are constantly attentive to the activities of the shark. Its bad reputation demands respect.

A pair of bull sharks near a reef show the grace and beauty of these large sharks.

during the period when females give birth. Such areas, often very murky, might reduce cannibalism by other larger sharks at a time when the newborn sharks are most vulnerable. There also may be less competition for resources because of the limited number of shark species that can tolerate waters with reduced salinity. Bull sharks are also frequently found in surf zones along the Atlantic coast of Florida as well as the Gulf coast. Because humans also prefer these locations, especially when waters are warm, there are frequent encounters with swimmers and surfers that often result in unfortunate interactions.

Bull sharks are also frequent inhabitants of the coral reef systems of southern Florida and the Bahamas. They are often confused with Caribbean reef sharks, much more common residents of coral reefs, or dusky sharks.

Size, Age, Growth, and Reproduction

Bull sharks reach a maximum length of nearly 10 feet (300 cm). Animals of this size are very stout and heavy bodied and appear even larger than their actual size. The International Game Fish Association (IGFA) lists a rod and reel world record bull shark of 8.7 feet (265 cm) with a weight of 697 pounds (316 kg), an exceptional weight for an animal of that length. Bull sharks grow about 7 inches per year (18 cm) for their first 5 years, 4 inches (10 cm) annually for the next 5 years, and finally at a rate of 2 inches per year (5 cm) for the rest of their lives. They may live for 25 to 30 years and perhaps even as long as 50 years.

Male bull sharks reach sexual maturity at around 72 inches (185 cm). Females reach maturity when they are a little larger, approximately 75 inches (190 cm). With their slow growth rates, these sizes mean that they are not be ready to mate until they are 10 to 15 years old.

Bull shark embryonic development occurs by way of a placental attachment to the mother. The gestation period is thought to be about 10 to 12 months. Bull shark litters range

The size and bulk of the bull shark explains why this species is such an effective predator.

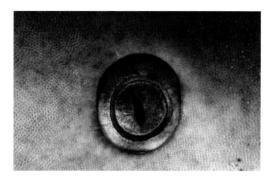

The beauty of the bull shark's eye belies its ability to detect motion, even under low light conditions, and helps to make it an even more effective predator.

from 1 to 12 pups and may have several fathers. The young are born at a length of 23 to 31 inches (60–80 cm). There has been speculation that many incidents involving bites from bull sharks occur in shallow water near shore when the females give birth. These favored nursery grounds are also locations where many swimmers may be present and encounters may be more frequent.

Food and Feeding

As might be expected, the bull shark is an accomplished predator. It feeds primarily on fish, but stomach contents have revealed a varied diet that may also include other sharks, stingrays, turtles, marine mammals such as dolphins, squid, shrimps, crabs, and occasionally birds.

Behavior and Interactions with Humans

Bull sharks are considered to be among the world's most dangerous sharks and are thought to be responsible for many attacks on humans. Their wide distribution and preferences for nearshore and inshore habitats may bring them into greater contact with humans and could explain the large number of attacks attributed to them. The International Shark Attack File lists 21 unprovoked attacks in Florida (21.4% of all attacks in which the shark's identity was confirmed). Two of these attacks were fatal.

Bull sharks have also been suspects in other attacks, most notably a series of attacks that occurred in New Jersey in 1916. Five attacks occurred in a 12-day period, resulting in four fatalities. Several of these occurred many miles up a tidal river. Such areas are more favored by bull sharks than most other species. The series of attacks is often blamed on a "rogue" white shark, but many experts cast doubt on the white shark's willingness to penetrate so far into freshwater; they prefer to blame the attacks on a bull shark. Both species are certainly capable of carrying out such attacks. However, there is not universal agreement on which species is more likely to have been responsible for these incidents.

Conservation and Management Status

Bull sharks are not considered to be a commercially valuable species, and most catches are incidental. If taken as bycatch, they are often kept for their fins, skin, liver, and meat. They are considered more valuable as a recreational fishery target, because of their size and weight. They are often on display in captive facilities because of their charismatic and ferocious appearance. The IUCN does not suggest a particular management strategy and lists them as "near threatened," a relatively low-priority category for conservation initiatives.

Caribbean Reef Shark

Common Names: Caribbean reef shark, reef shark

Scientific Name: *Carcharhinus perezi*

Identifying Features: Caribbean reef sharks are dark gray or grayish-brown with a lighter underside. The fins may be tipped with black, though not as deeply colored as in a blacktip shark. Caribbean reef sharks are frequently confused with other shark species, most often with bull sharks. Both have blunt and rounded heads. Bull sharks are generally more heavy-bodied, and their snouts are shorter. Novice divers who return from a trip to a local reef often report their escape from a near-death encounter with a bull shark, only to discover later that it was a Caribbean reef shark instead. Erring on the side of caution, however, is wise for new divers who lack experience diving with sharks. Experience helps one develop a sharper eye and a better sense of identification for shark species that may be of some concern. Bull sharks most certainly deserve a bit more vigilance than reef sharks.

Dusky sharks display some of the same characteristics as Caribbean reef sharks but have a snout that is more rounded. They are less frequent visitors to the reef systems preferred by reef sharks. Sandbar sharks also are sometimes victims of mistaken identity, though their broad, high, rounded dorsal fins and brown color distinguish them from reef sharks.

Frequency: common

Caribbean reef sharks are reminiscent of the yappy neighborhood dog that has to investigate everything that enters its domain. At the reefs near Big Pine Key, Florida, when a new dive boat arrives and disgorges its customers, a glance to the surface will—almost without fail—reveal a pack of snorkelers clustered together, all curiously swimming backward. On a closer look, one is likely to spot a reef shark

A quick glimpse of the snout end of the Caribbean reef shark shows its streamlined shape, blunt head, and black tips on the fins. These traits often confuse divers unfamiliar with these sharks and are responsible for their misidentification as bull sharks or blacktip sharks.

Seen in profile, this bulky Caribbean reef shark could easily be mistaken for a bull shark, its larger and more formidable cousin.

Commonly seen at the surface, Caribbean reef sharks may be the first residents to greet divers as they prepare for their underwater experience.

approaching the group and giving a toothy greeting. Once the guard shark is content with its initial visit, it resumes its leisurely patrol. It is an overly curious shark but easily satisfied once its initial reconnoiter is completed. Seldom are encounters more than a quick glance. Reef sharks' name comes from their frequent presence over and around coral reefs. They are streamlined and graceful

sharks, and their glides across the reef appear to be effortless.

Reef sharks are among the few species that can rest on the bottom; they don't need to constantly swim to breathe. Divers accustomed to encountering nurse sharks under ledges and in caves are often startled to spot a reef shark instead, and more than one diver has taken a hasty, extra-deep gulp of air when confused by such a surprise encounter. Famed shark biologist Eugenie Clark spent some time in the early 1970s researching what were called "sleeping sharks" and discovered groups of reef sharks in caves off Isla Mujeres near the Yucatan Peninsula resting quietly on the bottom with other species, including nurse and bull sharks. These behaviors had been reported a few years earlier, but they were little understood at the time and forced biologists to reexamine their thoughts about the basic biology of several shark species, including reef sharks. As common as this species is in Florida and the Bahamas, very little is known of its habits and biology.

Range, Distribution, and Habitat Preference

Caribbean reef sharks are commonly found on the reefs of southern Florida and the Bahamas. Though they are not restricted solely to reef habitats, they are not commonly encountered in the Gulf of Mexico or near the middle and northern east coasts of Florida, though they have been reported from the northern Gulf on occasion. They seem to prefer the warmer tropical waters of southern Florida and the Bahamas, where reef communities also thrive.

Studies of Bahamian populations of Caribbean reef sharks have indicated that, although they show some seasonal movement, when they are most abundant in the summer months, they seem to prefer specific areas. Their movements outside of this narrow range are thought to be limited, though they may forage to deeper waters on occasion.

They have been reported to reach depths as great as 1,167 feet (356 m).

Size and Reproduction

Reef sharks have been reported to reach lengths of 8.0 to 9.5 feet (250–295 cm), though the largest reported specimen has been subject to some skepticism; it may have been mistaken for another species. Sharks in the size range of 5 to 6 feet (150–185 cm) are more common. Males reach sexual maturity at lengths around 5.5 feet (165 cm), and females are thought to reach maturity at approximately 6.2 feet (190 cm). No studies have been conducted on age and growth rates, so there are no data regarding age at maturity.

Caribbean reef shark pups are born live. Females are thought to be pregnant for approximately 1 year and to reproduce every other year. Litter sizes are three to six pups,

Caribbean reef sharks, apparently curious, will often swim close to swimmers and divers when they enter the water.

and they are born at a length of nearly 2 feet (60–65 cm).

Food and Feeding

As might be expected from its preference for coral reef habitats, food items preferred by the reef shark include reef fish, flying fish, other small sharks, and some rays (yellow stingrays). Juvenile reef sharks may also feed on shrimps and other smaller crustaceans.

Caribbean reef sharks are common residents around the coral reefs of Florida and the Bahamas.

Caribbean reef sharks are often found in groups as they swim circuits around reef systems. They are common participants in managed shark dives and are readily attracted to baits in the water.

Caribbean reef sharks seem to glide effortlessly over coral gardens and are relatively easy sharks to photograph.

Behavior and Interactions with Humans

Reef sharks' preference for coral reefs mirrors the preference of most tropical divers. As a result, interactions are frequent and should be expected.

Among divers who seek deliberate encounters with sharks, baited dives have increased in popularity. Reef sharks are commonly encountered and are among the more reliable species to attend these free lunches. There are very few reports of problems caused by reef sharks, though they may display more frenzied feeding behaviors when large quantities of leftover bait are dispersed in the water following completion of a dive. Diving under those conditions would be riskier. The International Shark Attack File reports no attacks on humans by Caribbean reef sharks in Florida and only one unprovoked attack in the Bahamas; that attack was not fatal.

Conservation and Management Status

In most fisheries in other parts of their range, Caribbean reef sharks are taken as bycatch or by artisanal fishing efforts rather than as targeted species. Even so, studies of their ability to survive capture by long-lining methods indicate that they are strong and resilient and are likely to survive when released after bycatch.

The value of Caribbean reef sharks to reef systems as apex predators and their ecotourism value have led to protective measures in both Florida and the Bahamas, where harvest is prohibited.

The IUCN lists them as "near threatened" and suggests that better enforcement of fishing regulations, especially in marine protected areas, may be necessary to prevent the species from reaching a vulnerable stage.

Caribbean reef sharks often appear from nowhere, slipping over a coral outcropping, often startling an unwary diver.

Dusky Shark

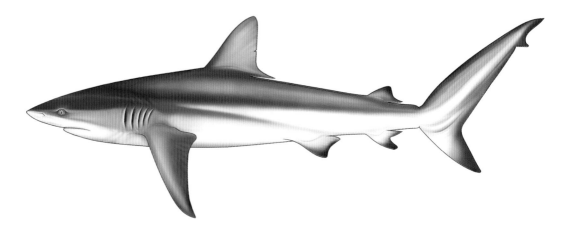

The dusky shark is comparatively unremarkable in its coloration. The grayish and subdued appearance of the pectoral fins and the dorsal surface explain its scientific species name "obscurus."

Common Name: dusky
Scientific Name: *Carcharhinus obscurus*
Identifying Features: The fins of dusky sharks are not well marked and are often termed "dusky," from which description they get their name. A dusky shark's snout is blunted but rounded. It has a large dorsal fin and relatively long, sickle-shaped pectoral fins. Dusky sharks are typically a gray or grayish-blue color with a whitish underside and may show a white stripe or band on the side near the tail. There is a ridge between the dorsal fins. These sharks are described as streamlined and slender, though larger specimens appear stout and bulky. The tips of the fins fade to a darker color, but the color is not well defined and crisp as in the blacktip shark.
Frequency: rare

Dusky sharks share some of the same reproductive traits as sandbar sharks, traits that first awakened scientists to the peril of population challenges faced by commercially valuable shark species. Both species take many years to replenish reduced numbers. Their slow rate of reproduction makes them ill-suited to survive an intense shark fishery. Yet both species are also highly prized for the value of their fins in the shark-fin-soup trade. This makes them a highly sought species and complicates management and conservation efforts.

Range, Distribution, and Habitat Preference

Dusky sharks are distributed worldwide and are common residents of Florida and the Bahamas. They show a preference for cooler waters but may be found inshore, perhaps

to give birth, and seaward to the continental shelf. They show no preference for freshwaters and are therefore not found in estuaries. They have been taken from depths as great as 1,300 feet (400 m). Their numbers have decreased because of their value to commercial fisheries, so they are not encountered as frequently as they were in the past. They are more likely to be caught by fishers than spotted by divers.

Size and Reproduction

Dusky sharks are reported to reach a maximum length of close to 13 feet (400 cm). Their average size is somewhat smaller, closer to 10 feet (320 cm). The dusky shark is one of the largest of the requiem sharks. The IGFA all-tackle world record dusky shark was caught off Longboat Key, Florida, and measured 9.8 feet (297 cm) in length and weighed 764 pounds (346 kg), a formidable record for a rod-and-reel fisher.

Their growth rate is extremely slow; dusky sharks may not reach sexual maturity for 20 years, when they have reached lengths of around 9 feet (280 cm). This ranks them as one of the slowest-growing shark species known. Pups are born live and are nourished by a placental attachment to the mother during embryonic development. Gestation is estimated to be around 18 months, though some studies predict 22 to 24 months. Females may reproduce at 3-year intervals or even longer. Litter sizes average 6 to 12 pups, though both smaller and larger litters have been observed. Pups are born at lengths of 35 to 39 inches (90–100 cm).

Food and Feeding

The dusky shark is a fast-swimming predator with a diet of many different fish species, including other fast-swimming fish such as tuna and mackerel. A wide variety of other fish have been found in dusky sharks' stomachs: herring, anchovies, groupers, many smaller fish, and a selection of rays and other species of sharks. Sea turtles and marine

The long pectoral fins and the rounded short snout are easily seen in this close-up of a curious dusky shark.

mammals have also been reported to have been meals for dusky sharks.

Interactions with Humans

Interactions with humans are rare. Underwater photographers report that dusky sharks are available on photo dives only when they are attracted by bait placed in the water to lure sharks to the dive site. Otherwise, only isolated encounters occur on routine dives. Their large size warrants caution when they are encountered, though the International Shark Attack File does not list any attacks in Florida or the Bahamas that were attributed to duskies.

Conservation and Management Status

It is no wonder that, with such a slow growth rate, such a late age at maturity, and such small litter sizes, this commercially valuable species has been decimated in heavily fished portions of its distribution. While the IUCN lists them as "vulnerable," not yet under worldwide threat, in the northeastern United States dusky sharks are highly regulated because of their diminished population. They are taken both by directed fisheries and as bycatch. When discovered as bycatch, dusky sharks are kept rather than returned to the sea because of the very high value of their fins. Initial regulations in US waters required that they be released by longline fisheries, and there are indications that some local populations are beginning to show signs of recovery, at least in terms of the numbers of juveniles that are observed. Adult populations remain low and will take decades to recover fully as juveniles grow to maturity.

Dusky sharks' highly migratory nature complicates management. Initiatives to regulate fishing in some countries may not be recognized when the sharks swim across country borders. Until management plans are adopted on a more global basis, species such as dusky sharks are at risk from unregulated fishing efforts.

Finetooth Shark

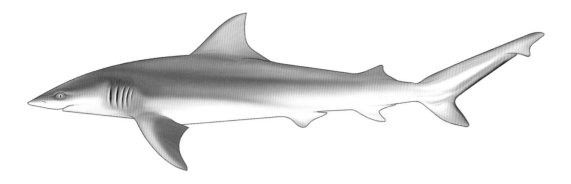

Common Name: finetooth shark
Scientific Name: *Carcharhinus isodon*
Identifying Features: Finetooth sharks are generally small but can reach lengths up to 6 feet (180–190 cm). Their snout is long and pointed. Their eyes are relatively large. They lack a ridge between the dorsal fins and are categorized as smoothback sharks. They are named for their needle-like tooth structure— not an easy way to identify them at a distance. The animals are light bluish-gray on the upper surface, fading to a white underside. They lack any distinctive markings on their fins. The dorsal fin is relatively large compared to other species.
Frequency: less common in southern Florida and the Bahamas, common in the Gulf

---

Finetooth sharks are common residents of very shallow inshore waters, preferring water depths less than 35 feet (10 m) in the summer. Thus they are found, often in large schools, near shorelines, bays, and estuaries where water temperatures are warmer. These shallow waters are also believed to serve as nursery grounds.

Range, Distribution, and Habitat Preference

The finetooth shark is well described from both Florida coasts, venturing well north in the Gulf of Mexico. Its southernmost range is thought to include the northern reaches of Florida Bay. This relatively recent observation provided evidence that the animals may actually move between the Atlantic and Gulf coasts. The finetooth shark has not been commonly found in the Florida Keys or in the Bahamas. Reports of its occurrence in Cuba have been disputed.

Size and Reproduction

Finetooth sharks are typically 4 to 5 feet (125–155 cm) in length, though a maximum size of 6 feet (190 cm) has been reported.

Finetooth sharks are aplacental viviparous sharks. Developing embryos are nourished by a yolk sac placenta and are born live. Females give birth to an average litter size of four pups after a gestation period of approximately 1 year. Females may reproduce every 2 years or at longer intervals in some locations but may reproduce annually in other locations. The reason for this difference has not been explained. Young are born at lengths of 18 to 21 inches (48–55 cm).

Male finetooth sharks are sexually mature at nearly 4 feet (125 cm) in length, and females are slightly larger, maturing at 4.5 feet (135–140 cm). At these sizes, males would be around 5 years old and females nearly 6 years old. Life expectancy is estimated to be 8 to 9 years for males and as long as 14 to 15 for females. Separate populations in different parts of the finetooth shark's range may mature at slightly different sizes and ages.

Food and Feeding

Croakers, Spanish mackerel, and mullet are preferred prey. Finetooth sharks are fast swimmers and will often feed on large schools of prey items such as menhaden as well. Juvenile Atlantic sharpnose sharks have also been reported in finetooth sharks' stomach contents.

Behavior and Interactions with Humans

Finetooth sharks have often been encountered in large schools. With the exception of their migratory behaviors, known to include movements into shallow waters during the summer months and back to deeper waters during the winter months, little is known of their behaviors. These movements may also include north-south patterns that place the animals in Florida during the winter and off the South Carolina coast in summer.

Few tagging studies have been conducted that have targeted finetooth sharks. Once such studies are undertaken, biologists will be able to better explain their migratory movement patterns.

There are no records of attacks in Florida, and it is likely that any injuries from finetooth sharks have come from improper or careless handling or during release of fish taken by recreational fishers.

Conservation and Management Status

Though finetooth sharks are not harvested in great numbers by commercial fisheries, they are often taken as bycatch in shrimp trawls. Their meat is considered to be tasty. They have been taken in gillnets, but because the use of those nets in coastal waters has been banned, that practice no longer presents a significant threat to finetooth populations. There is a small recreational fishery for finetooth sharks, but fishery managers do not believe it significantly impacts the sharks' populations.

Finetooth sharks are grouped with several other species, including the Atlantic sharpnose shark (*Rhizoprionodon terraenovae*) and the bonnethead shark (*Sphyrna tiburo*), into the "small coastal species" group for management purposes. In Florida there are no size limits for any of these species, though there are possession limits.

Finetooth sharks are considered a species of "least concern" by the IUCN. No recommendations have been necessary for managing their numbers. However, since their reproductive rate is very low, populations must be constantly monitored to prevent overfishing that could lead to a population collapse.

Great White Shark

Common Names: white shark, great white, white pointer, man eater, white death
Scientific Name: *Carcharodon carcharias*
Identifying Features: Great white sharks can reach enormous lengths, with girths that help to exaggerate their massive appearance. Their sheer size is one way to identify them. Their crescent-shaped tail and conical snout are equally important for identification. There is often a black spot at the posterior base of the pectoral fins. A lateral keel is present on the caudal peduncle (at the end of the body just before the base of the tail). Their slate-gray color and large black eyes are often noted by divers who have found themselves, by choice or by accident, in the water with these large sharks.
Frequency: rare in Florida and the Bahamas

There is absolutely no species of shark that has more mystique, more legends, or more fear associated with it than the great white shark. Peter Benchley's epic novel *Jaws* and the movie that followed the book raised the consciousness of the world to a point of universal fear and loathing for this sea monster. The fear of these animals led to some tourna-ments and all-out fishing efforts to eradicate them from the sea. Even more irrational behaviors occurred from this senseless fear. Shortly after the book and movie were released, I took a group of students to a fishing dock in Islamorada in the middle Keys to observe some of the catch offloaded from head boats and charter boats. It was almost always the case that sharks were among the catch, because prohibited species and trip limits had not yet appeared in the state or federal fishing regulations. We were amazed at the large sharks that were landed, none of which

The most classic view of great white sharks is the fin cutting through the surface of the water with ominous music playing in the background. These large predators have received more bad press than any other shark species.

Cruising just offshore, great white sharks are frequently encountered near shore where marine mammals, part of the preferred diet of great white sharks, are known to breed. Though Florida and the Bahamas lack these large marine mammals, great white sharks are still found in coastal and offshore waters.

The great white shark's conical snout, lunate tail, and the sharp break in color from a dark gray dorsal surface to a white underbelly are helpful aids in identifying them.

were great white sharks. We were equally amazed by the behavior of an elderly woman walking the dock who approached one of the dead sharks and began to beat the animal with an umbrella, her fear and hatred for these animals obvious from her behavior.

Mr. Benchley was equally disturbed by the public's response to his fictional tale, and he worked hard during his remaining years in conservations efforts. He was quoted in his later years in a piece for a traveling exhibit of the Smithsonian Institution's Ocean Planet, concluding that "the shark in an updated *Jaws* could not be the villain; it would have to be written as the victim; for, worldwide, sharks are much more the oppressed than the oppressors."

Great white sharks are clearly well adapted for their role as apex predators. Their size and large, well-equipped jaws certainly work in their favor. But their physiology also prepares them for this role because of their body's ability to regulate body temperature so that it is 7° to 9°F (4–5°C) above the ambient water temperature. This warming effect, the same principle used by athletes to prepare for physical exertion, adds a level of muscle efficiency that allows them to swim more rapidly and for longer durations than many prey items, giving them a hunting advantage over more traditionally cold-blooded fish.

Range, Distribution, and Habitat Preference

Great white sharks are worldwide in their distribution. In the Atlantic they reach north to Canada and south to the Caribbean. They have been recorded in Florida and the Bahamas and seem to prefer tropical, subtropical, and temperate waters. They are not considered to be common in the inshore waters around Florida and the Bahamas, but many underwater videos show them appearing from time to time on wrecks and reef dives in the Florida Keys, where they seem to react no differently than any other curious fish.

Commercial fishers, most notably swordfish long-liners, report that great white sharks are probably more common than most Florida chambers of commerce would have us believe. Some years ago, a good friend of mine, who fished commercially for swordfish, lured me to his boat on several evenings when I questioned his identification of animals he caught and released from his lines. He was polite in disregarding my skepticism and my questioning of his shark identification skills.

We fished for a week together, and I was captivated by some of the deep-water animals he recovered from the lines. Finally, on one recovery of the three-mile line, a much greater resistance was felt as the hydraulic drum recovered the line. John's comment to me was a suggestion that I be prepared to make an identification quickly, because he would immediately release the animal to prevent its being injured or killed. As the animal was reeled to the transom, there was simply no doubt regarding its identity. I was able to accurately estimate its size at 14 feet, and it remains the largest great white shark I have ever seen. Once the line was cut, the shark swam a lap or two around the stern of the boat, seemingly curious about the origin of the smells that it was sensing in the water. It then swam lazily away, apparently not bothered by its momentary capture. Later that summer, John was able to capture a Cuban night shark, rare for these waters, and it became a specimen in the collection of the Florida Museum of Natural History.

Size, Age, Growth, and Reproduction

Legends of white sharks reaching lengths of 25 to 30 feet (980–1,200 cm) have appeared in the popular press, where stories of lengths "bigger than my boat" are commonplace. Actual measurements, however, have seldom recorded lengths of more than 15 to 19 feet (5–6 m), still an enormous size for an animal regarded as one of the ocean's major apex predators. Exact measurements are difficult for animals of this size unless

Seen in a close-up view, the large black eye and conical snout of the great white shark are more recognizable.

The sleek and streamlined profile of the great white shark make it easy to understand why it is such a fast-swimming and agile shark, well equipped for its role as an apex predator.

they have been captured and can be measured on the docks. Actual measurements of specimens reaching 20 to 21 feet (6.0–6.1 m) have been taken, though maximum lengths are thought to exceed even these enormous sharks. Maximum lengths up to 23 feet (7 m) or longer may be possible. Animals of this size would easily exceed 4,000 pounds (more than 2,000 kg).

The problems in dependably locating great white sharks, coupled with the difficulty of working with such large animals, have complicated studies of age and growth. Accurately measuring animals that are tagged and released, then later recaptured and measured again, is the generally accepted method for determining growth rates and helping to determine aging. New techniques for managing captured animals, including specially designed lift platforms, and the widespread use of satellite tracking promise to close the gap in our understanding of great white shark age and growth. What little is known describes a slow-growing animal that takes many years to reach sexual maturity.

As popular as great white sharks have become, in-depth studies of their biology, and especially their reproductive biology, are not as complete as for other species. Few pregnant females have been captured for study, and much of the available information is based on estimates. A few studies predict that males must reach nearly 10 to 11 feet (3.1–3.4 m) before they reach sexual maturity and females must reach nearly 14 feet (450 cm). Ages at maturity have been revised and may be as great as 25 to 30 years. Some estimates place these sharks' life expectancy at 70 years or more.

Embryos develop internally but lack a placental attachment to the mother. A large yolk sac and consumption of eggs during embryonic development nourish the developing embryos. The great white shark's reproductive strategy results in litter sizes that may range from 4 to 14. The gestation period is unknown but estimated to be around 1 year. Pups are born at lengths of 43 to 63 inches (109–165 cm). Even at birth their size makes them worth attention.

This is probably not the preferred view of a great white shark. Its deep black eyes and fearsome tooth-filled mouth explain the fear that most swimmers, divers, and beachgoers have of these animals.

Food and Feeding

Because the great white shark is an apex predator of the highest order, almost any item in the sea could possibly serve as a food item for it. Marine mammals such as sea lions, sea otters, dolphins, other sharks, and a huge variety of fish species have been seen in stomach-content studies. Many video accounts show great white sharks actually jumping clear of the water in pursuit of sea lions or sea otters. Sea turtles and sea birds have also been recorded as prey items. Great white sharks are also frequently seen feeding on whale carcasses.

Behavior and Interactions with Humans

Other than their behaviors as consummate hunters, little is known of great white sharks' behavior. Most studies have concentrated on what shapes, colors, and prey items serve as the best attractants. Since marine mammals such as seals and sea lions make up much of their diet, it is not surprising that many studies of their hunting expertise have been detailed in areas where these mammals are found in large concentration. The shark's ability to jump completely out of the water in pursuit of sea lions has been popularized in the *Air Jaws* movies. Videos with stunning images of huge great white sharks leaping into the air with a victim clenched in their jaws are abundant. In addition, many surfboards have been recovered with sections removed by huge bites, perhaps a less enthralling demonstration of great white sharks' preference for whatever seems to resemble one of their favorite prey items.

Interactions with people begin with concerns about attacks. There is no question that great white sharks are potentially dangerous sharks. More attacks and fatalities are attributed to great white sharks than to any other shark species. Data from the International Shark Attack File, maintained by the Florida Museum of Natural History at the University of Florida, have shown that worldwide attacks by white sharks have increased over the past century. The researchers who maintain this file attribute this trend to several factors, but not to the idea that great white sharks have developed a taste for humans. They cite an increasing use of the water by people; better communication technology, which is now capable of obtaining information from remote parts of the world; and a wider coverage of suspected attacks by the media. While the number of attacks may have increased, the percentage of these that are fatal has decreased. This statistic is explained by advances in emergency medical care, including wider training of citizens and faster treatment for shock, often responsible for more deaths than the actual trauma from bites.

There have been many explanations for why great white sharks attack humans; several plausible theories have been suggested. The most popular explanation for many attacks is that humans swimming at the surface or paddling a surfboard appear from underwater to resemble marine mammals such as seals and otters, a highly preyed-upon group by great white sharks. Thus, some attacks may result from mistaken identity by great white sharks. Many other attacks have targeted spear fishers or commercial divers

The extreme bulk, well-developed sensory systems, powerful jaws, and remarkable swimming speed and agility position great white sharks as one of the ocean's most effective and accomplished predators.

who are spearing fish or collecting abalones, lobsters, or other attractive food species. The odors, sounds, and sights of potential food items attract great white sharks and result in attacks originally directed to the items taken by the diver.

While there is no real directed commercial fishery for great white sharks, they are often represented in incidental bycatch. In previous years they became sought-after trophy fish, and many were captured simply for their jaws and bragging rights. Great white sharks experience a great deal of capture stress, and the mortality of animals captured on longlines or by trawl is thought to be high.

Most of the record fish have been taken from Australian waters. The IGFA all-tackle world record measured 16.8 feet (513 cm) and weighed a staggering 2,664 pounds (1,208 kg). The animal was landed by rod and reel with a 130-pound test line.

Conservation and Management Status

Though great white sharks show the capability for long-distance movements, satellite tracking has shown that they also return to favored areas during certain seasons. Their wanderings have challenged scientists to understand why they go where they go. Whether they are following food sources, seeking mates, or traveling to nursery grounds is simply not yet well understood. Because a better understanding of these movements is essential to reach adequate conservation and management decisions, many studies are attempting to answer these questions. Accurate measurements of population sizes have not been possible; this lack further complicates management decisions.

Recognition of the ecosystem value of great white sharks as apex predators in marine ecosystems has led to wide-scale protective measures around the world. Australia and South Africa, areas that are known to support large populations of great white sharks, have implemented restrictions on interactions with them. California and Florida also have strict protective measures, which have been extended all around the United States.

The great white shark is listed as "vulnerable" by the IUCN, largely based on the lack of knowledge about this species. While the animals can be common in some portions of their range and are an immensely popular species of shark, they are scarce enough that little is known of the details of their life histories.

Great white sharks have long been a prized target of recreational fishers. In recent years they have become extremely valuable as a recreational dive encounter. Cage-diving expeditions in locations where the animals can be found with some predictability have become a growing ecotourism industry and have exerted some pressure to keep populations intact and healthy. Great white sharks are yet another example of species whose commercial worth may be greater for live animals than those sacrificed for food or for commercial by-products.

Hammerhead Sharks

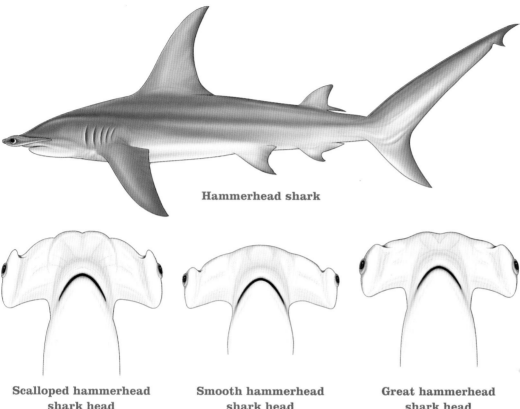

Hammerhead shark

Scalloped hammerhead
shark head

Smooth hammerhead
shark head

Great hammerhead
shark head

Common Names: hammerhead, great hammerhead, scalloped hammerhead, smooth hammerhead (bonnethead; see separate account on bonnethead shark)

Scientific Names: *Sphyrna mokarran* (great hammerhead); *Sphyrna lewini* (scalloped hammerhead); *Sphyrna zygaena* (smooth hammerhead); *Sphyrna tiburo* (bonnethead)

Identifying Features: Identification of the three largest species (great, smooth, and scalloped hammerhead sharks) is often complicated at a distance. While they are easily recognized as hammerhead sharks, determining the differences between them is difficult unless they can be observed at close range. The scalloped hammerhead shark is perhaps most easily recognized. There is an indentation in the middle of the front of the head and two more indenta-

tions, one on each side of the midpoint, giving the head a scalloped appearance. The smooth hammerhead shark lacks these indentations, and the front of its head has a more straightened appearance. The great hammerhead shark, by contrast, has only a shallow notch in the very midpoint of the hammer but is otherwise smooth in appearance. Differences in the size and placement of the dorsal, pelvic, and anal fins are additional keys to telling them apart. Unfortunately, these finer points of identification are more easily determined from animals that are dead. The traits have less value for identifying living animals that are swimming in the shallows or on a reef, or fighting on the end of a fishing line. Even when they are captured, mistakes are common. Some well-respected taxonomists say that the only real way to be absolutely certain of identification is through DNA analysis, which is impractical and unimportant for divers, fishers, or boaters, who are content to know they've seen a hammerhead shark!

Frequency: common

There are probably no other sharks in Florida or Bahamian waters, or even worldwide for

Hammerhead sharks are the most easily recognized shark in Florida and Bahamian waters. The three largest species, the great hammerhead shark, the smooth hammerhead shark, and the scalloped hammerhead shark, are not easy for the average person to tell apart unless samples from all three are side by side. The great hammerhead shark is the largest of the species.

The smooth hammerhead shark does not have the same indentations as the great hammerhead shark and the scalloped hammerhead shark. Commercial and recreational catches of hammerhead sharks indicate that the smooth hammerhead shark may not be present in the Gulf of Mexico or the Bahamas.

that matter, that are easier to identify than hammerhead sharks. Their uniquely shaped head readily distinguishes this group of sharks from any other. Anyone who sees a hammerhead shark for the very first time asks the same question: what in the world happened during evolution to produce such a strange-looking shark? Young kids who see one of these sharks brought to the dock have a ready explanation: it was run over by a truck and its head got smashed. While not acceptable as a biological explanation, such a simple conclusion has a certain appeal.

The shape of the head seems to offer a number of advantages to hammerhead sharks that set them apart from other species. The head is often referred to by biologists as a cephalofoil (*cephalo* is Latin for head or skull, and a "foil" is a surface that can assist in providing lift). An airplane's wing (airfoil) and the struts on a specially designed boat called a hydrofoil both aid in achieving lift. In the same way, the cephalofoil structure of the hammerhead shark may function much as the bow planes of a submarine do. The lift that the head provides helps the shark to move up and down in the water much more easily and with more control than a shark with a more traditionally shaped fish head can. Anyone who has seen a hammerhead shark chase a prey item is amazed at how fast and effective these sharks are. The design of the head presumably helps to make this possible.

In addition to aiding swimming movements, the wide head places the eyes and the nostrils (olfactory nares) farther apart than in other species. This greater separation means the shark can detect odors over a wider scent trail (odor corridor). It may even be able to cover a wider area with its vision. In addition, the large size of the head allows for more electro-sensory pores to be spaced over the head, compared to other shark species. More sensory pores theoretically improve hammerhead sharks' ability to detect the weak electrical currents produced by the nerves and muscles of living prey. The combination of these sensory modifications greatly enhances the hammerhead shark's hunting skills and makes it a formidable hunter.

Fishers who seek the great game species of Florida and the Bahamas, especially permit and tarpon, have seen firsthand how effective hammerhead sharks are as hunters. Stories at the docks include tales of huge tarpon being bitten in half by hammerhead sharks that appeared from nowhere and attacked a hooked fish. This spectacle often occurs in water so shallow that it seems impossible for such a large shark to even find enough room to swim. The first sign of trouble is usually the sight of a gigantic dorsal fin cutting through the water. The fin seems disproportionately tall for the size of the shark and often looks more like the sail of some small sailboat racing over the flats. Fishing guides coax their customers to reel a hooked fish to the boat as fast as possible if a hammerhead shark is nearby, in hopes that the fish can be released and escape a charging hammerhead shark before the helpless fish becomes a meal.

At least four species of hammerhead sharks are present in Florida and Bahamian waters. The bonnethead shark is the smallest of the four, attaining a maximum size of about 60 inches (152 cm). Because its habits and distribution differ significantly from the other three species and its classification is a bit different, it is discussed separately in an account of its own.

Range, Distribution, and Habitat Preference

All three species are found worldwide. They tend to inhabit waters inside the continental shelf and the near coastal waters, especially in temperate and tropical seas like those of Florida and the Bahamas. At various times of the year, especially in the spring, hammerhead sharks may invade bays and estuaries. They may be using these areas as prime hunting grounds or mating or nursery grounds, though very little is known about where hammerhead sharks mate or give birth.

The scalloped hammerhead shark has multiple indentations on the snout and, in a mature specimen viewed at close range, is relatively easy to distinguish from the other species.

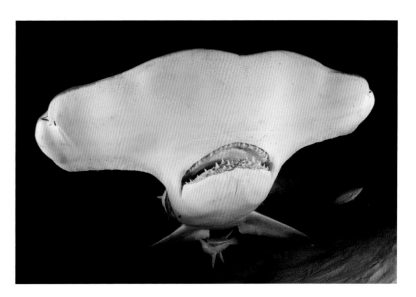

This view of the cephalofoil, the hammer of the hammerhead shark, illustrates the width of the head and the wide separation of the eyes and the olfactory nares (nostrils) of the hammerhead shark. Also visible are the huge number of pores, the ampullae of Lorenzini, the external openings of the electrosensory system. With such a wide head, and with the mouth placed well under the body, the ability to detect the weak electric currents of a living organism at close range when the hammerhead shark makes it final approach may be crucial to making an effective attack.

Fishers in Florida are especially aware of the presence of hammerhead sharks near areas where tarpon are caught. As spring begins and the water starts to heat up, small fish, followed by large fish, start to appear

The hammerhead shark's bizarre appearance is not enhanced when its mouth is open.

Perhaps the best description for this picture of photographer Jill Morris is to borrow a line from the movie *Jaws:* "Swim, Charlie, swim. Take my word for it. Don't look back."

inshore. Following the small fish are larger fish, including many prized gamefish such as bonefish, permit, and tarpon. Not far behind them are hammerhead sharks, which can ruin a good day of fishing by taking chunks from even very large tarpon. Their speed and agility often force fishers to reel a fish as fast as they can and hope the gamefish can be released in time to escape a charging hammer. It is exciting and a little frightening to see a 10-foot hammerhead shark roaring across

water less than 3 feet deep, pushing a huge wake in front of it, to grab a fleeing tarpon.

The great hammerhead shark is more abundant than the scalloped or the smooth hammerhead shark in Florida and Bahamian waters. There is some evidence that the smooth hammerhead shark is rare or absent in the Gulf of Mexico and Bahamian waters but more common in northeastern Florida. Most reports of hammerhead sharks that were once caught either commercially or by recreational fishers simply describe the catch as a hammerhead shark. Without more extensive measurements, perhaps even DNA analysis, absolutely certain identification may not exist for these catches, and distribution is therefore not completely understood.

Size, Age, Growth, and Reproduction

Hammerhead sharks can reach lengths of more than 10 feet (305 cm), and specimens of more than 15 feet (4.6 m) have been encountered. Some accounts even suggest lengths of 19 to 20 feet (6 m), though detailed measurements for such specimens apparently don't exist. The great and scalloped hammerhead sharks may not even reach sexual maturity until they are more than 8 feet (245 cm) in length. The scalloped hammerhead shark may reach maturity earlier in its life cycle at the length of 6 feet (185 cm), though it can grow to more than 13 feet (395 cm).

Very little work has been done to study age and growth in these species. Some estimates suggest that they may live more than 30 years, perhaps even more than 55 years for the scalloped hammerhead shark. There is little knowledge of annual growth rates. Since hammerhead sharks are seldom kept in captivity, growth estimates come from studies of tagged animals. Such studies rely on accurate length measurements when an animal is captured, and they depend on the animal surviving after being tagged and then being recaptured at some later time, with the same attention given to accurate measure-

As the hammerhead shark swims close to the photographer, its haunting eye never seems to leave the diver, creating a very eerie encounter.

ments. Few such studies exist for hammerhead sharks. The few studies that have been successful estimate that growth rates may be up to 12 inches per year (32 cm). Rates are thought to be greater than that for juveniles and much slower later in the shark's life when growth slows, as it does for most animals.

After fertilization, hammerhead shark embryos are attached to their mothers by a placenta, from which they obtain nutrients during embryonic development. Once they mature, they are born alive. Once the pups are born, they no longer receive any maternal care. Fishers often witness spontaneous births when female hammerhead sharks are caught near the end of their pups' embryonic development. This happens especially with the smallest of the hammerhead sharks, the bonnethead shark, and often the young will still have remnants of the placental cord still attached. The stress from capture may explain such spontaneous abortions witnessed by fishers.

The gestation period for hammerhead

Hammerhead sharks, like most species of sharks, are effective hunters at dawn and dusk because of the low-light adaptations of their eyes that permit detection of prey when the light fades. Although commonly seen on the seafloor hunting for stingrays, great hammerhead sharks utilize the entire water column when other prey species are available.

sharks is thought to be between 10 and 12 months. Litter sizes may range from 20 to more than 30, and size at birth can be from 20 to 30 inches (50–80 cm). Females may reproduce every other year or at even longer intervals.

Food and Feeding

Hammerhead sharks are fast-swimming, aggressive feeders. They feed on fish of all species but seem to have a special appetite for stingrays. Many underwater photographs and videos have been taken of hammerhead sharks pursuing rays. The shark's ability to follow the ray, even though it may bury itself

Changing seasons can be measured in many different ways. One sure sign that the water temperatures are beginning to warm is the appearance of hammerhead sharks in the shallow waters near shore.

Hammerhead sharks are the main attraction for many dive experiences in the Bahamas. Even a biologist could be made to wonder if this hammerhead shark is actually smiling for the cameras and visitors or preparing for a meal.

under the sand, shows that it is an effective hunter and is easily capable of feeding on rays that may be 4 to 5 feet in diameter. Stomach contents of hammerhead sharks caught during tournaments have revealed many stingray barbs as well many different species of fish and small sharks. Scalloped hammerhead sharks, in some areas of their distribution, seem to feed almost exclusively on squid.

Behavior and Interactions with Humans

Feeding is certainly the most often observed behavior of hammerhead sharks. But most of their lives is spent in secrecy. One exception is the schooling behaviors of hammerhead sharks that have been observed. Huge aggregations of hammerhead sharks have been frequently observed, more often in the Pacific near underwater seamounts than in Atlantic and Caribbean waters. Scientists are uncertain what prompts these aggregations to form. Many animals show scars resembling those that often, in various species of sharks, occur during times of mating, but very few observations of actual mating by hammerhead sharks have been reported. Newborn sharks are generally not seen in these areas, so there is little evidence to suggest that the animals congregate near a nursery grounds. The most often-cited explanation of the aggregations points to group feeding activities, where animals may congregate in a small area and then move away to forage for food, returning after successful hunting. It could also be a behavior that serves as a prelude to migration as a school, though more research will be required to establish for certain why these large aggregations occur.

Though hammerhead sharks are often large and are often found in areas frequented by divers and beachgoers, attacks attributed to hammerhead sharks are rare. The International Shark Attack File reports only 13 unprovoked attacks in Florida (13.3% of all

A common sight in the shallow, clear waters of the Bahamas and southern Florida is a great hammerhead shark making the rounds. Stingrays are common residents of the sandy area around reefs and are a favorite food of hammerhead sharks.

the attacks in which the shark's identity was confirmed) when bites from hammerhead sharks have been confirmed, and no fatalities have been recorded. No attacks attributed to hammerhead sharks have been recorded in the Bahamas.

In recent years, as hammerhead sharks' migratory patterns have become more understood, divers flock to areas where hammerhead sharks are present to dive with them. Some areas of the Bahamas have thus become shark destinations, and the ecotourism industry has expanded to serve this interest. The sharks are docile, and the interactions with divers and photographers have resulted in extraordinary photographic images and videos as well as a better understanding of these animals.

Because the animals are easily stressed when captured, few hammerhead sharks have been captured for public display, and even then, only the largest of the public aquariums are able to provide adequate space for these animals.

Conservation and Management Status

Hammerhead shark populations worldwide have experienced sufficient declines that many have been assessed as endangered (scalloped and great hammerhead sharks) or vulnerable (smooth hammerhead sharks) by the IUCN. The IUCN lacks the power of en-

In spite of its unusual appearance, the hammerhead shark in all its grace is still a wonder of nature and an awe-inspiring sight.

forcement; it merely evaluates data and recommends policy changes. All three species, however, have been listed internationally by the Convention on International Trade in Endangered Species (CITES), which does provide some level of protection. The US Endangered Species Act also lists the scalloped hammerhead shark as an endangered species.

More importantly for Florida and Bahamian waters, Florida law prevents the landing of hammerhead sharks in state waters and mandates immediate release, unharmed, if one is caught while fishing. This is best interpreted and described in an article by David Shiffman, who noted that

> if a fish is brought out of the water, it is "landed." If anglers stop the act of releasing a fish to measure it or take a photo, it is not "immediately released." If a fish isn't "immediately returned alive and unharmed" (and if the extremely physiologically stressful act of bringing a hammerhead shark out of the water results in it dying after release, it was not released "unharmed"), it is harvested. If you drag the shark out of the water and leave it there until it stops moving long enough that you feel safe to approach it, that is not an "immediately released" animal, and it isn't an animal that is "released unharmed." Landing and/or harvesting hammerhead sharks is illegal. (Shiffman 2016)

The Bahamas have taken protection to new levels by banning all shark fishing—not just fishing for hammerhead sharks—in Bahamian waters.

Lemon Shark

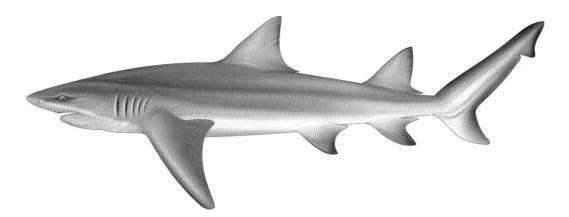

Common Names: lemon shark, sand shark
Scientific Name: *Negaprion brevirostris*
Identifying Features: Lemon sharks are generally light-colored, most commonly a yellowish brown, which gives them their name. This coloration provides some protection in the shallow sandy areas, where young lemon sharks blend in well. Their presence in this habitat may explain why they are often incorrectly referred to as sand sharks. The young may have black fin tips, causing them to be commonly confused with small blacktip sharks. However, the lemon shark's first and second dorsal fins are approximately the same size, and this feature easily distinguishes them from blacktip sharks, whose second dorsal fin is much smaller than their first dorsal fin.
Frequency: very common

Lemon sharks have been studied in more complete detail in natural populations than probably any other species of shark. Pioneering studies by Dr. Samuel ("Doc") Gruber of the University of Miami and the Bimini Biological Field Station have examined almost every aspect of the physiology and life history of this shark and have led the way for similar studies of other species.

The lemon shark is one of the several species of sharks that do not need to continually swim to breathe. This feature makes them ideally suited for laboratory studies. Beginning with laboratory studies of vision, the research has expanded to study every facet of lemon shark biology, from feeding and metabolism, to age, growth, migration, and reproduction. At a lab based in Bimini in the western Bahamas, hundreds of students and volunteers have worked to help under-

The lemon shark is one of the most common inshore sharks and can be found over grass beds, in the shallow bays and mangrove forests, in sand flats, and on the reefs—almost anywhere around the coasts of Florida and the Bahamas.

Common sharks and voracious feeders, lemon sharks are often hooked by fishers. The sharks will often break off the line and may be left with hooks and tackle in their mouths. The remnants of a hook can just be seen in the left side of the lower jaw of this large lemon shark.

stand the role of lemon sharks in the tropical marine ecosystem.

Range, Distribution, and Habitat Preference

Lemon sharks are very common inshore sharks in Florida and the Bahamas. They prefer the more tropical warmer waters, though they are found along the East Coast as far north as New Jersey and as far south as Brazil. They are frequently found in shallow inshore waters of Florida and the Bahamas, on coral reefs, and in bays and estuaries, though they do not seem to have a preference for less saline waters. Public aquariums often feature lemon sharks because of their charis-

Lemon sharks are often solitary animals but are also frequently found in groups. Whether they are a social species, like nurse sharks, or whether they gather for feeding or mating is a matter for continuing studies.

The shallow waters are not just homes for neonates and juvenile lemon sharks. It is not uncommon for full-sized adults to enter these waters, especially during the time when adult females give birth.

matic appearance and their ability to tolerate life in a captive facility.

They are known to give birth to their litters inshore, where the young may spend their early years, and they are very common as juveniles in shallow inshore bays around seagrass beds and the coastal mangrove forests of southern Florida, the Florida Keys, and the Bahamas. Prior to their protection in Florida waters, smaller lemon sharks were often taken for food; their flesh is highly regarded by many fishers. They are commonly caught by sport fishers who are targeting bonefish, permit, and tarpon, since they inhabit the same waters as these more highly prized game species.

Size, Age, Growth, and Reproduction

Lemon sharks may reach a length of 9 to 10 feet (290 cm), though lemon sharks of this length are considered unusually large. Corresponding weights of animals of this size would be nearly 400 pounds (180 kg). One specimen, however, was reportedly measured at a length of 12 feet (368 cm), suggesting that larger sharks of this species may exist. Larger ones have been reported from estimates, but actual measurements of these lemon sharks have not been provided. Males are believed to become sexually mature at lengths around 7 feet (220 cm), and females mature when they reach 7.5 feet (230 cm). Reproducing animals are between 11 and 13 years old. They may live longer than 20 years.

The shallow inshore bays and estuaries serve as nursery grounds for lemon sharks. Neonates and juvenile lemon sharks such as this small animal can be seen in the mangroves, where they often share these habitats with other small sharks, especially nurse sharks and bonnethead sharks.

No species seems to be immune to the loving embrace of remoras. These symbionts apparently benefit from food scraps left over from feeding lemon sharks and provide a parasite cleaning service to their hosts.

Growth rates of lemon sharks are low and quite variable. Results from studies in Bimini, Bahamas, show rates of less than 2 inches (5 cm) per year during their first year, while populations in Brazil and the Marquesas Islands in the Florida Keys grow at faster rates, nearly 7 inches (18 cm) per year.

Lemon shark females reproduce at two-year intervals. Young are born alive, and there is a placental attachment to the mother during embryonic development. Females are pregnant for 10 to 12 months and give birth to litters of 12 pups on average. Newborns are between 20 and 26 inches (55–65 cm). An individual litter may have several fathers.

Nursery grounds in the Bahamas have been well characterized. Young lemon sharks are born in the shallow mangrove-sheltered bays and may reside there for several years before they begin to forage at greater distances from shore. Similar nursery areas have been found in southern Florida, the Florida Keys, and the Ten Thousand Islands on Florida's west coast as well.

Food and Feeding

The primary food source of lemon sharks is fish. Grunts, parrotfish, snappers, porgies, and mojarras and pinfish are commonly found in their stomach contents, and many other fish species have also been consumed.

Shrimps and crabs have also been discovered. Small sharks, including smaller lemon sharks, and rays are also taken. This range of prey items shows that lemon sharks are effective predators and are able to survive on whatever food source might be readily available. They are also known to be effective nighttime feeders. Behavioral studies have revealed that lemon shark feeding may also be temperature-dependent.

Behavior and Interactions with Humans

In what are regarded as some of the earliest groundbreaking shark behavioral studies in 1959, Dr. Eugenie Clark trained two lemon sharks to ring a bell for food. Clark's study indicated the sharks' ability to learn and, perhaps equally important, to remember the tasks that had been learned. She also discovered that as the water temperature cooled during the fall, the animals' feeding rate slowed, and when the water temperature dropped below 68°F (20°C), the animals ceased to feed. After nearly 10 weeks, the water temperature rose above 68°F (20°C), and the sharks resumed their feeding.

Lemon sharks can reach large sizes and are accomplished hunters. They are formidable fighters when caught by rod and reel. The IGFA lists the all-tackle world record lemon shark at a length of 94 inches (239 cm), with a weight of 405 pounds (183.7 kg). Animals reaching this size would deserve attention and respect if only for their size. But in addition, fishers often report that releasing large lemon sharks is complicated by what is best described as a serious attitude problem; care must be taken to avoid injury at the side of the boat during release, because of the shark's resistance to capture.

The International Shark Attack File does not list any attacks attributed to lemon sharks in Florida. However, four unprovoked attacks were reported in the Bahamas. None were fatal.

Lemon sharks are one of a few species of sharks that does not have to swim to breathe. It may rest motionless on the bottom and continue to respire effectively.

Conservation and Management Status

Because of the extensive research in the shallow lagoons of Bimini, Bahamas, scientists have come to understand the vital role of inshore habitats in the early life history of some coastal sharks, such as the lemon shark. Development in these islands has threatened to seriously affect and change the shallow waters to support a larger human population. Studies from the Bimini Biological Field Station and the University of Miami have shown that shallow inshore bay areas throughout the Bahamas and southern Florida serve as nursery grounds for lemon sharks. Divers and fishers are very likely to see these small sharks, especially around the grass beds and mangroves that are so common in these areas. Snorkelers who explore the submerged mangrove root systems will almost always encounter juvenile lemon sharks, who begin their lives in these rich shallow bay areas. Studies have shown that juvenile lemon sharks remain residents of these areas for many years and that adults will return to the same areas, perhaps for mating or pupping. Encroachment into these areas by expanded coastal urban development provides an additional threat to already challenged populations.

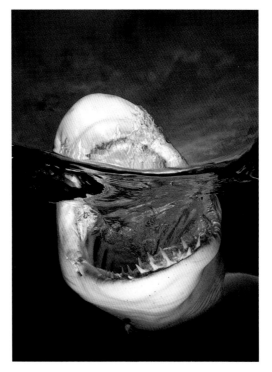

An effective nocturnal feeder, the lemon shark was one of the first to have its vision systematically studied. It was shown to be better adapted for nocturnal vision than many of its prey species.

The IUCN reports a large market for lemon shark flesh; the fins are also considered valuable. Because there is not a very intense worldwide fishery for lemon sharks, they are listed as only "near threatened." Local regulations in some areas, including Florida, prohibit harvest in state waters by recreational fishers, to protect locally dwindling population numbers.

Mako Shark

Common Names: mako, shortfin mako
Scientific Name: *Isurus oxyrinchus*
Identifying Features: The mako shark's deep blue coloration becomes sharply white on the underside, and its long pointed, conical snout and half-moon (lunate) tail make this shark easy to distinguish from other species. In addition there is a ridge, termed a keel, on each side of the animal, just forward of the tail fin. These structures are called lateral keels, and that segment of the tail is referred to as the caudal peduncle; these terms will be useful if readers seek other texts to make positive identification.
Frequency: rare

The classic pose of the mako shark displays its very pointed, conical snout and the pointed teeth that characterize this ocean predator. Among the fastest fish in the sea, the mako shark is a spectacular acrobat and an accomplished hunter.

I know of no one—recreational fisher, commercial fisher, diver, or boater—who encounters a mako shark who isn't genuinely excited by the experience. Mako sharks are gorgeous animals. Their metallic blue coloration and streamlined design are the essence of perfection in design and capability. Sport fishers who happen to hook a mako shark are in for the fight of their lives. A frequent jumper and a fish capable of stripping all the line from a reel on its initial run, the mako shark was historically one of the few shark species listed as a game species by the IGFA.

Mako sharks are commonly considered to be among the fastest fish in the seas, swimming at speeds of up to 25 miles (40 km) per hour for short durations and capable of short bursts of more than 40 miles (64 km) per hour. One possible explanation for such extraordinary capabilities is their unusual ability to maintain their body temperature at levels 7° to 9°F (4–5°C) above the ambient water temperature, a feature they share with white sharks. Muscle activity is generally

more efficient at higher body temperatures. Human athletes recognize the value gained from warming up before exercise. Some shark species that are highly regarded as fast-swimming, effective predators have elevated body temperatures as well, and mako sharks are foremost among these species. Most bony fish and most species of sharks do not have this ability.

Range, Distribution, and Habitat Preference

Shortfin mako sharks (usually referred to in this book just as mako sharks) are found worldwide in temperate and tropical waters. They prefer cooler water, though they have been taken from southern Florida and the Bahamas, generally from deeper, offshore waters. Some sources suggest that they can be found as deep as 1,600 to 1,950 feet (500–600 m). One study using satellite tracking technology showed that the sharks could attain depths of 2,900 feet (888 m), where water temperatures could be as low as 39°F (4.6°C). They are frequently confused with longfin mako sharks (*Isurus paucas*). Side-by-side comparison easily eliminates confusion, since the length of the pectoral fins is very different between the species. Longfin mako sharks are also found worldwide but are generally not common in Florida and the Bahamas and are found more frequently at large distances offshore and in deeper waters than shortfin mako sharks.

Mako sharks are highly migratory and capable of very-long-distance movements. Distances of up to 2,452 miles (3,923 km) have been recorded from the time a shark was tagged to when it was recaptured. Where it might have gone between those times is unknown, so the actual distance traveled is likely to have been much greater. Satellite tagging programs are starting to reveal more about the fine-scale movements of sharks than tag–release–recapture studies can reveal. Some of these movements have also been rapid. One satellite-tracked mako shark

reportedly traveled 90.6 miles (145 km) in 45.4 hours.

Size and Reproduction

Mako sharks can reach lengths of 12 to 13 feet (365–400 cm) and weights in excess of 1,000 pounds. The IGFA world record is 1,221 pounds (555 kg), though one specimen reportedly captured by a scientist weighed 1,249 pounds (567 kg). Lengths are perhaps better estimates of size, since weights fluctuate considerably depending on whether an animal has recently fed or was in a starved condition when weighed. Mako sharks are voracious feeders, and a large meal could add considerably to its recorded weight.

The mako shark is a large pelagic shark that does not survive well in captivity, so little is known of its reproduction. Nursery grounds and preferred mating sites have not been described. Males may reach sexual maturity at lengths of 6.5 feet (200 cm), and females are thought to mature at much larger sizes, perhaps as long as 9.5 feet (290 cm). These lengths suggest that males mature at an age of 7 to 9 years and females at approximately 21 years. Because it is a commercially valuable species, this slow growth to maturity means animals have to survive fishing pressures for a long time to replenish their numbers. Maximum life expectancy may be nearly 30 years or more.

The bluish color of the dorsal surface fading to a white underside are distinctive color markings of mako sharks. The lunate tail with its nearly equal upper and lower lobes also aids in identification. Though sleek in appearance, mako shark bodies are also stout and muscular.

The arsenal of teeth and gaping jaw coupled with the mako shark's speed and agility make it an effective predator. Photographer Andy Murch described this posture as a threat display, shown when he approached the shark too closely. When he backed away from the shark, its mouth would close and other postural changes and vibrating head displays would cease. Whenever he swam closer to the shark, the display would resume—a warning that should not be taken lightly.

Mako sharks are known to give aplacental live birth to anywhere from 4 to 30 young at a time, though somewhere between 10 and 20 seems to be more common. Embryos feed on eggs that continue to be produced by the female during their development but rarely consume their siblings.

No reliable estimate of the gestation period exists. Ranges of 6 months to 2 years have been reported, with most agreeing that 15 to 18 months is most likely. Pups range in size from 25 to 28 inches (65–72 cm). Since most shark females wait for some period of time after giving birth before mating again, it seems likely that female mako sharks may be reproductively active at three-year intervals or longer.

Food and Feeding

Stomach-content studies have shown that mako sharks feed on a wide variety of fish, including bluefish, other sharks, and swordfish. Squid, turtles, and marine mammals have also been discovered, though probably more as a result of opportunistic feeding than

as preferred prey. Numerous injuries from swordfish have been seen in mako sharks, suggesting that swordfish may be a frequent target, though bluefish seem to be more preferred where their range overlaps with feeding mako sharks. Some estimates report that mako sharks may consume up to 4.6% of their body weight per day to meet their metabolic demands. This would suggest that a 100-pound (45 kg) mako shark would require 4.6 pounds (2 kg) of food every day. These same studies calculate that a mako shark feeding predominantly on bluefish could theoretically devour nearly 1,100 pounds (500 kg) of bluefish per year.

Behavior and Interactions with Humans

In Florida waters only one unprovoked, confirmed attack by a shortfin mako shark has been recorded by the International Shark Attack File. It was not fatal. No attacks have been reported from Bahamian waters. Since mako sharks are generally encountered more commonly offshore and in deeper water, the likelihood of encounters with divers or swimmers is low. They are much more likely to be encountered by fishers, and injuries during capture or release by fishers have been noted.

With the availability of high-quality underwater camera and video systems, the mako shark has become a prized photographic subject. Ecotourism dives where mako sharks are more frequently encountered than the shallower waters of Florida and the Bahamas frequently seek mako sharks for their customers. As a result, some astounding photographs have been produced, and some details of the sharks' display behaviors are becoming better understood.

Conservation and Management Status

The mako shark is prized for its meat, perhaps more than any other shark species. While it is targeted for meat and fins, it is also a significant bycatch in the swordfish and tuna fisheries. Its worldwide population is thought to be in decline, as much as 40% by some estimates, and the IUCN regards it as a "vulnerable" species. It is a regulated fishery in the United States, and its quota is closely monitored. However, since it is a highly migratory pelagic species, protection outside US waters is not assured. Its slow growth, late maturity, and three-year reproductive cycle do not prepare the species well to survive an intense commercial or recreational fishery.

Nurse Shark

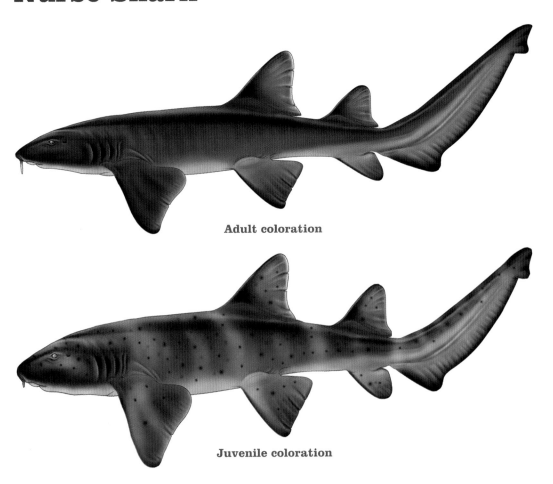

Adult coloration

Juvenile coloration

Common Names: nurse shark, gata, gato
Scientific Name: *Ginglymostoma cirratum*
Identifying Features: The nurse shark can be distinguished from other Florida and Bahamian sharks by the presence of two fleshy barbels, appendages found on the snout that resemble the whiskers of a catfish. The barbels are located adjacent to the nostrils (olfactory nares), just above the mouth. They are thought to help the shark locate prey that may be hiding just under the surface of the bottom sediments or sands.

In most species of sharks, the first dorsal fin is significantly larger than the second dorsal fin. In nurse sharks, however, both dorsal fins are large, rounded, and prominent

and may be nearly the same size, a feature they share with lemon sharks, sandtiger sharks, and sawfish. Their tail (caudal fin) lacks a lower prominent lobe and extends beyond the body as much as one-quarter of the total length of the animal. The color is generally tan or light brown but may range to a deep brown or even dark gray coloration.

Frequency: very common

The very first shark I ever caught was a nurse shark. The north jetty of the St. Johns River in Jacksonville was a wonderful place for young kids to fish in the early sixties. Redfish, shads, drums, and assorted bottom fish were

Nurse sharks are one of the most common inshore sharks in Florida and the Bahamas. They are found in virtually every habitat, from reefs to river mouths, bays, estuaries, and shallow bays and marshlands.

plentiful and tested our horrible knot-tying skills. Bait was obtained by pulling our small boats up to a shrimper, passing aboard a washtub with one dollar inside it, and being rewarded with a tub full of the shrimpers' bycatch to be used for bait. The strange fish caught in the shrimpers' trawl nets amused us while we waited for a fish to take our baits and offered hours of curious observation.

We mainly fished for redfish and king mackerel. They were edible and satisfied our parents that our fishing was actually valuable in providing food for the table. Our tackle was pathetic, the sort that broke teenagers could cobble together from pieces and parts. One summer morning was particularly productive, with five or six respectable reds in the cooler before noon, even though I had lost several bottom rigs on what I believed to be large rocks or pieces of debris on the bottom. One of the rocks actually seemed to give way a bit, and I struggled to lift it off the bottom. It seemed to be coming slowly to the surface, and I wondered if I had hooked a log, or a large boot, or a body (teenagers have active imaginations). Very suddenly it broke the surface, slapped its tail against the side of the boat, catching two of us in the face, and

snapped the line. We knew what it was instantly, and we also knew that our sad excuse for fishing tackle was no match for a 6-foot nurse shark that was more like a boulder with fins.

Catching a shark was exhilarating, and we soon after began to target our fishing efforts to catching more sharks. On the rare occasion when we hoisted one to the surface, almost always a nurse shark, we were mostly beaten to death by the struggling shark, which simply did not like being caught. Nothing has

Long considered to be a social species, nurse sharks are frequently found in groups. Animals within a group may represent all size ranges. During and after mating season, groups may be sexually segregated, with large numbers of females gathering together.

The group behavior of nurse sharks can easily be seen even by a casual observer. A boater walking down a dock in the early morning for a day on the water might be a bit surprised to be greeted by an unconventional welcoming committee when the tides are exceptionally high.

changed since those early years. Nurse shark attitudes have not been refined at all; nurse sharks continue to destroy my gear and abuse my crews, even though I now have better gear and more experienced assistants.

Nurse sharks are among the most common nearshore sharks in Florida and the Bahamas. Compared to other sharks, they are relatively slow and are generally bottom-dwellers. As would be expected, measurements reveal that their metabolic rates are among the lowest determined for any shark species.

While most sharks must continually swim to breathe, nurse sharks possess the musculature that allows them to draw water across their gills to breathe. This makes it possible for them to rest on the bottom. They are well suited for captive studies and for aquarium displays, and almost all large aquariums that have shark exhibits feature nurse sharks.

Nurse sharks are one of the most popular species for basic research, because they are so easy to keep in captivity. Studies that require a shark to be immobilized, such as studies of vision, or smell, or hearing, or brain function, can use nurse sharks with a minimum of special restraints to keep them from swimming away. Special equipment to pump water across the gills to prevent suffocation is not needed unless the animal is deeply anesthetized in the study.

Range, Distribution, and Habitat Preference

Nurse sharks are found throughout the tropical and temperate waters of the western Atlantic, the Caribbean, and the Gulf of Mexico. Their range has been reported to extend north of the Carolinas and south to Brazil and to tropical regions of West Africa in the eastern Atlantic. They are commonly encountered in nearshore and inshore waters throughout Florida and are present on all reef systems in southern Florida, the Florida Keys, and the Bahamas. They are commonly found under ledges, near rock outcroppings, around wrecks, and in reef systems. They are also found in shallow bays and estuaries and are very common in mangrove habitats.

Size, Age, Growth, and Reproduction

Size estimates for nurse sharks vary widely. They have been reported to reach lengths of 14 feet (more than 4 m). Actual measurements from captured specimens show that their maximum size is probably much less than that, perhaps up to 11 feet (335 cm). The exaggerated sizes reported by fishers usually come from estimates ("it was as long as my boat") or from the magnification effects that occur underwater and can increase apparent sizes reported by divers by 25–30%. As they grow in length and increase in weight, their girth and overall bulk may also make them appear to be larger. Commercial shark fishers who catch nurse sharks as unintentional by-catch are generally not happy when a nurse shark turns up on their line. Some fishers state that catching a large nurse shark is like trying to pull the plug on the ocean bottom because of the shark's reluctance to give in to a fight and its desire to hold on to the bottom!

Tagging studies with nurse sharks that have been measured, tagged and released, and later recaptured have shown that they grow very slowly, averaging about 5 inches (13 cm) per year. Newborn animals may grow slightly faster, up to 12 inches (30 cm) in the first year, 10 inches (27 cm) in each of their second and third years, and then a more standard rate of 4 to 6 inches (10–15 cm) per year until their growth stops in old age. Male nurse sharks reach sexual maturity at a length of around 82 to 86 inches (210–220 cm) and are probably 18 years old. Females are a little larger before they mate, between 86 and 94 inches (220–240 cm), and would be about 20 years old.

Nurse sharks are the only species of shark whose reproduction and reproductive behavior has been studied extensively and systematically in the wild. A common sight in the shallow waters of southern Florida, the Bahamas, and throughout the Caribbean, mating nurse sharks have been noticed by boaters, divers, fishers, and casual beachgoers for decades. Their preference for shallow waters near shore provides easy opportunities to view them in large aggregations during the summer months. Their mating behaviors are visible from long distances because their tails can often be seen slapping the water, frequently less than 2 to 3 feet deep. The large size of the animals may help make their mating visible from such distances.

In fact, it was this behavior that first attracted me to investigate and begin a long-term study of reproduction with my colleague Wes Pratt, then of the National Marine Fisheries Service. As a scientist whose main interest was the biology of nurse sharks, I had no clear understanding of the reproduction in this species. Pratt's strength was his understanding of shark reproduction in pelagic species such as blue sharks, mako sharks, and great white sharks. Yet he had never seen sharks mating, something that still is true of most shark biologists, even now. I invited him to join me in the Florida Keys to see what we might learn, and our project blossomed into more than 20 years of research, in a project that Wes continues.

Wes's experience with shark tagging allowed us to tag and recognize individual nurse sharks and to conclude that they return to the same general area to mate, most males every year and females every 2 years.

Females appear to be active for 2 to 3 weeks during their reproductive season and will mate with different males. Further studies with animals that we captured after mating, with the professional assistance of Sea-World Adventure Parks, showed that nurse shark litters, averaging anywhere between 20 and 40 pups, may have as many as four to six fathers. Nurse sharks show little migration,

Because shallow inshore areas are frequented by nurse sharks, these sharks are probably the ones most commonly sighted by anyone on or near the water. They are common residents in boat basins and near waterfront dock areas, where they often wait for scraps from the fish-cleaning table. Juvenile nurse sharks most frequently inhabit the near-shore areas; there they may escape predation pressures. Once they reach sexual maturity, they tend to move into deeper waters offshore.

The color patterns of neonate and juvenile nurse sharks are very different from those of adults. The spots and bars that are present in young animals disappear as they age and become a more uniform brown to dark gray as adults.

There is no explanation for what might have attracted this nurse shark to the camera housing. Nurse sharks seem to have no reluctance to pose for the camera and are seldom considered a threat to divers unless the diver happens to be collecting lobsters or other morsels of food that nurse sharks might find appetizing.

so multiple fathers provide a way to ensure some genetic diversity rather than having 40 baby sharks with the same father. Sharks that were captured for this study gave birth after 4.5 to 5.0 months. These animals were all returned to the study site after their year of captivity, and the females returned in subsequent years to mate again.

Food and Feeding

Nurse sharks prefer to eat crustaceans such as crabs, lobsters, and shrimps but will also feed on fish, albeit pretty slow fish. They are suction feeders and can exert enough suction force to pull a conch from its shell or an octopus from small cracks in the reef. A narrator for a cable television show, trying to relate to the viewers of the show, once claimed that "nurse sharks have the combined sucking power of twelve Hoover vacuum cleaners," an impressive fact I was never able to confirm. Nevertheless, the nurse shark is known to generate the greatest suction force of any aquatic feeding vertebrate.

When nurse sharks inhale a food item at the surface, the sound of the suction force is very evident. It is this noise that is often cited as the origin of the term "nurse" to describe this species. A more likely origin, researched by Dr. Jose Castro, suggests that the term "nusse," originally used around 1440 to describe large fish, eventually became "nurse" in the seventeenth century to describe the modern-day nurse shark.

Divers who seek lobsters can attest to the nurse shark's appetite for these spiny creatures. Many divers' bags full of lobsters have been stolen by nurse sharks. Divers reluctant to give up their catch have unfortunately become victims of stray bites from nurse sharks seeking a free meal.

Behavior and Interactions with Humans

Because nurse sharks are so common, many different behaviors have been identified, most from casual observations rather than from systematic studies. Their social behavior is well known. They are frequently found in large aggregations resting under ledges or docks or in groups on grass beds or in mangrove habitats.

A common sight above the reef, in caves, or under ledges, nurse sharks are drawn to any habitat that offers structure or cover. They are seldom seen without the company of remoras. Even when they are caught by rod and reel, remoras will often accompany the nurse shark into the boat.

Early experiments that evaluated the ability of nurse sharks to learn showed promising results. Eugenie Clark's feeding experiments in the late 1950s revealed their ability to be trained to respond to a bell for a food reward and showed that they could remember the task after their training had been interrupted for several months. In more novel experiments, the US Navy conducted experiments that trained nurse sharks to retrieve rings from a training pool. The sharks were trained to find the rings and return them to the trainer. Why such experiments were undertaken was not revealed in the study.

If nurse sharks aren't the most commonly encountered shark in Florida and Bahamian waters, they are close to the top of the list. They are taken frequently by fishers who are fishing bottom structures (reefs, rock piles, artificial reefs, etc.) and, less commonly, by trolling. Game fishers in southern Florida and the Bahamas also find them on the flats when fishing for Florida game fish such as bonefish, permit, and tarpon. Divers encounter nurse sharks in almost every habitat.

They are less likely to be spotted in heavy surf areas.

They are considered by most divers to be docile and nonthreatening. In fact, as senseless as it may seem, divers have been known to grab nurse sharks by the dorsal fin to catch a ride. Many of these adventures have ended poorly for the diver. Also, grabbing a small nurse shark by the tail often results in a nasty bite and the discovery that once a nurse shark has bitten, it seldom releases its prey. There have been numerous newspaper articles reporting on swimmers or divers who visited an emergency department with a shark still attached to a hand or an arm. Larger bites have more catastrophic results; they can lead to massive tissue loss from the strong, crushing jaws and deceptively sharp teeth of nurse sharks.

The International Shark Attack File lists seven unprovoked attacks in Florida (7.1% of all the attacks in which the shark's identity was confirmed). None of these attacks were fatal. No attacks attributed to nurse sharks have been reported in the Bahamas. It is highly likely, however, that many provoked incidences have occurred in which injuries were probably minor and no report was filed.

Conservation and Management Status

Although they are taken in some fisheries, nurse sharks are not highly prized as a commercially valuable species. They have an extraordinary ability to destroy fishing gear, especially when brought to the side of a boat, and most commercial fishers prefer to release them quickly. Though they survive well following capture, death from bycatch is considered to be a potential threat.

Some conservation actions have been taken locally in some countries in the Caribbean, but populations of nurse sharks are generally regarded as stable and not under any particular threat. The IUCN lists them in the category "data deficient" in most parts of their distribution, though they are considered to be vulnerable in some South American countries where fishing pressures are more intensive.

Oceanic Whitetip Shark

Common Names: oceanic whitetip, whitetip
Scientific Name: *Carcharhinus longimanus*
Identifying Features: The white coloration of the tips of most fins is the most distinctive identifying feature of oceanic whitetip sharks, though the color may fade as the animal ages. The tips of the second dorsal fin and the anal fin may be black-tipped rather than displaying the white coloration of the other fins. The shark's dorsal and pectoral fins are disproportionately large and well rounded. Oceanic whitetip sharks are also very heavy-bodied and stocky with a large, blunt snout. The body color is grayish to gray-brown with a lighter underside. A black patch, often referred to as a saddle, may be present on the dorsal surface near the second dorsal fin. And there is a ridge on the dorsal surface between the first and second dorsal fins.
Frequency: rare nearshore, more common offshore

Oceanic whitetip sharks are a common open-water species of shark. Some estimates suggest they may have once been among the top three most abundant shark species of all oceanic sharks. Current estimates indicate that their numbers have decreased substantially in recent years because of the high value of their fins to the shark-fin soup commercial fishery. Oceanic whitetip sharks are not to be confused with the reef whitetip shark (*Triaenodon obseus*), a much smaller species found in the Indo-Pacific. Reef whitetip sharks, as the name suggests, prefer a different habitat and are most commonly found on, under, or near shallow coral reefs, often in large huddled masses, rather than in the open deeper water preferred by oceanic whitetip sharks.

The large size, greatly rounded dorsal and pectoral fins, and the broad white markings on the fin tips serve to help identify oceanic whitetip sharks. This female has what appear to be numerous bite wounds, suggesting she may have recently been involved in mating behaviors.

Range, Distribution, and Habitat Preference

Oceanic whitetip sharks are distributed worldwide, mostly in tropical or warm temperate water. They prefer deeper open waters and are seldom encountered near shore, the exception being islands that may be surrounded by very deep waters. In spite of their preference for the open ocean, satellite tracking of tagged whitetip sharks from the Bahamas has shown that they will actually return faithfully to the same area and may have stronger ties to particular regions in the oceans than was previously known. These same studies have also shown that oceanic whitetip sharks dive somewhat deeper than was previously thought, reaching depths at night as great as 3,550 feet (1,082 m). Scientists studying these movements have suggested that the whitetip sharks were following the movements of large billfish and tuna that are known to congregate in these areas. The sharks' long-distance movements may therefore be driven by changes in food sources and food prey movements.

Size and Reproduction

Oceanic whitetip sharks average about 6.0 to 6.5 feet (185–200 cm) in length. Their stout body at these lengths makes them appear much larger. There are reports of animals reaching a maximum length of 11.5 feet (350 cm), though that is generally regarded as exceptionally large for this species. Oceanic whitetip sharks are more commonly measured at lengths closer to 6 to 7 feet (185–215 cm). In fact, the IGFA lists its all-tackle world record oceanic whitetip shark at a length of 7.2 feet (220 cm) and a weight of 369 pounds (167 kg), from San Salvador in the Bahamas.

Few data exist describing the reproduction of oceanic whitetip sharks. They are thought to become sexually mature at a length of about 5.5 to 6.0 feet (170–180 cm) and an age of 4 to 5 years. Embryos are nourished by a placenta and are born live following a gestation period estimated to last 1 year. Litter sizes range from 1 to 15 pups, averaging around 10, and pups are approximately 25 to 29 inches (55–75 cm) in length when they are born.

Food and Feeding

Food preferences include open-ocean pelagic fish species such as tuna and dolphinfish (mahimahi). Oceanic whitetip sharks may also feed on squid and may actually follow pilot whales as they effectively hunt for squid as one of their primary food sources. Stingrays, turtles, marlins, swordfish, and seabirds have also been reported from stomach contents. Feeding frenzies have been frequently observed, suggesting some need for caution in managed, baited diving encounters. Andy Murch described some challenging moments that occurred while he took photographs (included here) of oceanic whitetip sharks:

> A picture of an oceanic whitetip [was] taken at Elphinestone Reef [in the] Red Sea. . . . This was one of five that have been hanging round this reef for several months now. This one is not . . . pleased at my presence. Unfortunately, some idiots on a live aboard were dangling a chicken on a rope, teasing [the shark] up near the surface and when it saw me, [it] viewed me as a threat to its meal. After [receiving] a good push into my camera and a second pass pushing

Oceanic whitetip sharks have reportedly shown agonistic displays, often arching their backs, opening their mouths wide, and pointing their pectoral fins. Moving away from the shark will usually result in ending the display posture.

An open-water shark that is found at great depths, perhaps chasing pelagic prey from deeper water, individual oceanic whitetip sharks often return to the same areas, suggesting some site fidelity or philopatry.

into my shoulder, I just gently sank down a few meters and it seemed to accept this as me backing down, and it returned to its chicken.

Interactions with Humans

The International Shark Attack File does not list any attacks by oceanic whitetip sharks in Florida, probably because of their preference for open waters and avoidance of nearshore waters where divers and beachgoers might be present. Fishers are far more likely to encounter them in deeper offshore waters of southern Florida and the Bahamas. They are often observed in the company of silky sharks (*Carcharhinus falciformis*). Nevertheless, oceanic whitetip sharks have been described as aggressive and potentially dangerous. Numerous accounts blame them for attacks and casualties associated with shipwrecks and airplane disasters.

Conservation and Management Status

Worldwide, the oceanic whitetip shark is listed as "vulnerable" by the IUCN. How-

The distinctive white tips of the oceanic whitetip shark make it easy to identify when it approaches boats or is viewed by divers underwater.

ever, extensive fishing pressures in some portions of its range have severely reduced its numbers and have led to a local classification of "critically endangered," particularly in the northwestern Atlantic and the western central Atlantic. The IUCN notes that Gulf of Mexico populations may have decreased by as much as 98% to 99% over the past 40 years, certainly a cause for concern in this part of the world. This dramatic change demonstrates how local pressures can deplete a population while numbers in other regions may remain more stable. The recent satellite tracking studies showing that oceanic whitetip sharks will return to select areas may mean that local regulations, in addition to international management, might need to be made more stringent to protect these animals.

Sandbar Shark

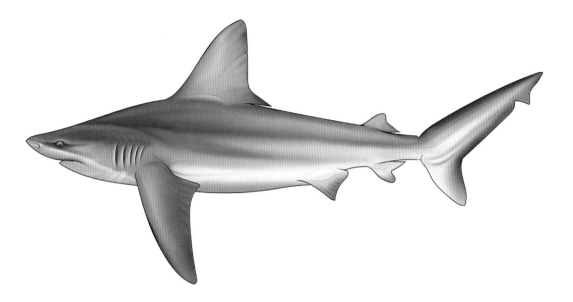

Common Names: sandbar, brown shark
Scientific Name: *Carcharhinus plumbeus*
Identifying Features: The brown color of the upper body of the sandbar shark and its high first dorsal fin are very distinctive. It also has a whitish underside, and its pectoral fins are very long compared to other species. It is a very heavy-bodied shark, among the largest of the coastal species.
Frequency: very common; most abundant species on the East Coast of the United States.

If you have ever eaten shark meat, there is a high probability that it was a sandbar shark. Although mako sharks and blacktip sharks are also consumed in large quantities, the sandbar shark has long been one of the sharks most highly sought after by commercial fisheries. In fact, the fishery was so intense at one time that sandbar shark populations were reduced to very low levels; many scientists and fishery managers feared their populations might never recover. Sandbar sharks are well represented in the recreational fisheries as well. Early estimates placed the sandbar shark second only to blue sharks in the recreational fishery. Strict management of the fisheries has begun to improve population stability and has led to partial recovery in some areas.

Range, Distribution, and Habitat Preference

Sandbar sharks are widely distributed in tropical and temperate seas and are very

The brown coloration and tall dorsal fin make the sandbar shark an easily recognized species as it passes through Florida and Bahamian waters. *Photograph courtesy Terri Roberts. Used with permission. All rights reserved.*

common around Florida and the Bahamas, where they are found in shallow coastal waters and in bays and inlets. They do not show a preference for low-salinity waters, so their appearance nearshore is limited to more coastal regions. Though sandbar sharks have been taken from deeper waters, their preference seems to be for shallower, nearshore habitats over sandy or muddy bottoms. Such preferences may help to explain the origin of their common name. Large populations have been studied in Delaware and Chesapeake Bays.

Size, Age, Growth, and Reproduction

Sandbar sharks grow to a length of nearly 8 feet (240–250 cm), though 5 to 6 feet (155–185 cm) is more common. Many studies have examined aging and growth because of the importance of these data for management of overfished populations. Studies have shown that growth rates early in a sandbar shark's life may be around 8 inches (22 cm) per year and that its growth may drop to 4 to 7 inches (11–19 cm) per year for 2- to 5-year-old animals. Thereafter the growth rate may stabilize at 1 to 2 inches (2–4 cm) per year. This is an extremely slow rate of growth.

Because of the sandbar shark's importance to the commercial fishery and the perceived drop in abundance from fishing pressure, originally estimated to be as much as 80% to 90% in a 10-year span, the reproductive biology of sandbar sharks was extensively studied to determine how rapidly its diminishing numbers could be naturally replenished.

The information revealed by these studies showed that sandbar sharks have a reproductive strategy that does not serve them well as a targeted species. Females are capable of bearing young every 2 to 3 years. In order to reproduce, however, they must reach lengths of approximately 6 feet (180–190 cm). This translates to an age of 8 to 14 years for males and 8 to 16 years for females. There are estimates suggesting that even these ages may be

understated, that growth to maturity could take as long as 30 years and that the shark's life expectancy may therefore be considerably longer than that.

Since sandbar sharks' growth rates are very slow and they reach maturity relatively late in their lives, they must survive the intense fishery in some parts of their range in order to reach reproductive age. In addition to slow growth and late maturity, their litter sizes are relatively small, averaging 6 to 10 pups, though published accounts vary widely. Up to 85% of their litters may be sired by more than one male. However, some investigations have indicated two to as many as five fathers for the several litters examined. The shark's natural rates of population replenishment are therefore very low, another trait that places population stability at risk. These reproductive characteristics explain why there is such great concern for the welfare of populations that have been reduced by fishing pressures. Their ability to rebound, even if fishing was stopped entirely, would likely take decades.

Sandbar sharks are aplacental viviparous, and the females give birth to their pups in shallow inshore nursery grounds. Pups range in size from 20 to 27 inches (50–70 cm). Most nursery grounds on the East Coast of the United States are concentrated in northern waters rather than Florida or Bahamian waters. Bay waters off New Jersey, the Delaware and Chesapeake Bays, and other similar waters seem to be preferred, though there are reports of newborn pups from as far south as Cape Canaveral. Pregnant females have also been found in the northern Gulf of Mexico, leading biologists to assume that nursery grounds are also located in these waters.

Food and Feeding

Sandbar sharks prefer a diet of fish. Their stomach contents include a wide variety of fish: eels, blue runners, snappers, small sharks and rays, mollusks such as squid and octopuses, and shrimps and crabs. They have

Sandbar sharks are also known as brown sharks; the coloration is evident in this photograph of a captive sandbar shark. These animals' sharklike appearance makes them a valuable species for large display aquariums, where they survive well.

been characterized by some biologists and fishers as opportunistic bottom-feeders, consuming almost any food item that might be available.

Behavior and Interactions with Humans

Sandbar sharks survive well in captivity. They are a favorite species for captive facilities because they are more "sharklike" in appearance than other species like the nurse shark, which also does well in display aquariums.

There are four recorded instances of unprovoked attacks by sandbar sharks on humans in Florida and one record of an unprovoked attack in the Bahamas. All were nonfatal. Sandbar sharks tend to avoid beaches and surf zones and prefer sandy or muddy bottoms rather than coral reefs, so their chances for encountering a human are relatively slim. Because of their potentially large size, however, they deserve respect if encountered while people are diving or swimming.

The large dorsal fin and long pectoral fins are on full display in this oncoming sandbar shark. *Photograph courtesy Terri Roberts. Used with permission. All rights reserved.*

Conservation and Management Status

For many years the sandbar shark has been known to be a major component of commercial fisheries as a targeted species. Its meat is prized, and its large fins are valuable to the shark-fin trade. It is taken by longline, bottom nets, and hook and line. Recreational fishers also harvest significant numbers by rod and reel: it is considered a prized game fish among shark fishers. Because sandbar sharks have a very slow rate of growth, a late onset of sexual maturity, and low numbers of offspring, they are particularly vulnerable to overfishing. The IUCN has classified the sandbar as "vulnerable," the category just below "endangered," because of the threat faced by populations worldwide from commercial and recreational pressures. The sandbar shark was one of the first species to be studied in such detail, and the conclusions from these studies set the standard from which recommendations for other species have been made.

Sandtiger Shark

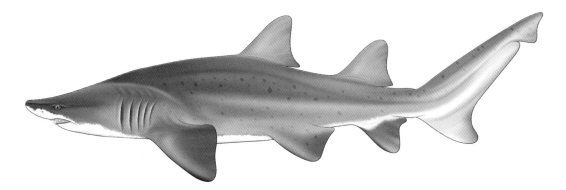

Common Names: sandtiger, sand shark, ragged tooth (Australia), gray nurse shark

Scientific Name: *Carcharias taurus*

Identifying Features: The most distinctive feature of the sandtiger shark is the shape of its head and mouth and the prominent display of teeth. The snout is long, bluntly pointed, and slightly flattened. The mouth seldom closes completely and leaves many rows of teeth exposed. The two dorsal fins are almost the same size. Sandtiger sharks are not easily confused with other Florida and Bahamian species. In spite of their name, they are not closely related to tiger sharks (*Galeocerdo cuvier*).

Frequency: less common

If you could select just one species of shark to display in a public aquarium, it would most likely be a sandtiger shark. Sandtiger sharks are a particular favorite because of the structure of their mouth and jaw that exposes a huge number of long, pointy teeth. This makes them appear to be menacing and gives the fearsome appearance that visitors apparently expect to see when they visit facilities with tanks large enough to hold bigger sharks. In the case of sandtiger sharks, however, the look is deceiving. They are generally considered to be relatively docile and nonthreatening and usually swim slowly and gracefully.

Range, Distribution, and Habitat Preference

Sandtiger sharks are nearly worldwide in distribution, preferring nearshore temperate

Its heavy body and large, tooth-filled mouth gives the sandtiger shark a more ferocious appearance than it deserves.

Sandtiger sharks are large, heavy-bodied sharks. Their pointed snout, dorsal fins of almost equal size, and a mouth that is always open, prominently displaying their teeth, are keys to their identification.

This profile reveals the wide, tooth-filled jaws that characterize sandtiger sharks.

imagine descending to a shipwreck in 90 to 100 feet (27–30 m) of water, penetrating to the interior of the wreck, and coming face-to-face with a jaw full of pointed teeth staring at you through the gloom. The encounters always end well, for both shark and diver, and most of the best underwater photographs of sandtiger sharks come from such encounters. While caught occasionally by fishers in southern Florida, the Bahamas, and the Gulf of Mexico, these sharks are more common in the Atlantic waters of northeastern Florida.

Size, Age, Growth, and Reproduction

Sandtiger sharks can reach nearly 10 feet in length (300–310 cm). These larger animals can weigh as much 330 to 350 pounds (150–160 kg), though there is a report of an animal that may have weighed a staggering 500 pounds (220–230 kg). These larger specimens are very bulky and deep-bodied.

Male sandbar sharks are thought to reach sexual maturity at around 75 inches (190 cm) and 6 to 7 years old. Females become mature at a length of 87 inches (220 cm), around 9 to 10 years old. Females are thought to reproduce every other year or at longer intervals. Usually they give birth to just one or two offspring—definitely one of the lower rates of reproduction for sharks. Life spans are

and tropical waters. They are most common in waters less than 600 feet (191 m) deep and prefer rocky bottoms near rubble or caves. In some parts of their range, they are encountered in nearshore surf zones; perhaps this explains why they are commonly referred to as sand sharks in some areas.

Sandtiger sharks are very commonly seen or caught near shipwrecks. If you are a diver, largely unknown, though some animals held in captivity have lived 30 years or more.

Sandtiger sharks show some of the more unusual and interesting aspects of shark reproduction. What makes them unusual is what happens during embryonic development. There is no placental connection to the mother. Nourishment must come from some food source, and embryonic sandtiger sharks apparently consume their littermates as one of their major food sources; this is one of very few species that demonstrate embryophagy ("-phagy" means eating, consuming, or feeding upon). Female sandtiger sharks continue to produce eggs for long periods after mating. These eggs, fertile or unfertile, eventually make their way into the uteri, where they are also consumed by the developing embryos, in a process termed "oophagy," which is more common in sharks than embryophagy. If two embryos survive to birth, it is because only one embryo in each of the two uteri was able to survive the cannibalism that occurs within each uterus. The combination of intrauterine cannibalism and oophagy is not known to be widespread among sharks, but it is certainly an effective way to nourish developing sharks.

In theory, such a feeding strategy, which can produce larger offspring at birth, is especially useful for sharks that give birth in shallow waters near shore, such as sandtiger sharks, because in those locations there might be more competition and a greater abundance of potential predators. Closely related species, such as mako sharks, also consume eggs during embryonic growth, but they generally give birth in more open, deeper waters, where competition might be less. This may explain why their offspring are smaller at birth. While such an embryonic feeding strategy for sandtiger sharks may be an effective way to nourish growing embryos, it is not one that will allow for large numbers of offspring; sandtiger shark populations can easily suffer long-term damage if population declines result from severe fishing pressure.

Food and Feeding

To reach their large sizes and weights, sandtiger sharks must consume a great deal of food. They feed mostly on fish of nearly

Sandtiger sharks are common residents of shipwrecks, often startling divers who may be exploring inside sunken vessels. Their presence around wrecks has been exploited by some commercial fishers, who have completely eliminated populations of these sharks in some areas. These sharks are slow to reproduce, so their populations require a significant time to replenish their numbers.

Captive facilities delight visitors with sandtiger sharks because of their fearsome appearance. This shark is a prized display animal and can be found in almost all captive facilities. It adapts relatively well to a captive environment.

Sandtiger sharks are not considered dangerous, in spite of their frightening appearance. They are relatively slow-swimming sharks with no history of attacks on humans in Florida or the Bahamas.

Behavior and Interactions with Humans

Most sharks depend on oils stored in the liver to manage buoyancy. Otherwise, they must either rest on the bottom or swim constantly in order to change depth. Sandtiger sharks have developed a slightly different approach to achieving buoyancy control. They come to the surface and gulp air, storing it in their stomach to maintain neutral buoyancy. By this strategy, they can remain motionless in the water without either having to swim or having to rest on the bottom. Occasional burping will void the air from the stomach when the animal has a need to change its position. Some aquarium workers have found sandtiger sharks upside down at the surface of the water with very distended bodies. They concluded that too much air was swallowed and the animal became so positively buoyant that it floated to the surface and inverted. Animals found in this condition are turned upright and eventually achieve more normal buoyancy and balance. Whether the same thing occurs in the wild is not known.

There have been occasional reports of groups of sandtiger sharks surrounding schools of fish and cooperating in group feeding behaviors. Some of the first descriptions of this type of behavior came in 1915 from Dr. Russell J. Coles, one of the associates of Ocean Leather Company, a firm that pioneered the production of high-grade leather from shark skin. He was speaking of observations at Cape Lookout.

every species. Studies of stomach contents of captured sandtiger sharks have found small sharks as well. Average weights of large sharks who feed primarily on large fish are difficult to estimate and are dependent on what they have eaten and how recently they have fed. Fluctuations in reported weights are therefore not surprising.

This shark works in [a] more systematic way in securing its food than any shark of which [I] know. On one occasion [I] saw [a] school of [one] hundred or more surround [a] school of blue-fish and force them into solid mass in shallow water, and then at the same instant the entire school of sharks dashed in on the blue-fish. On another occasion with [a] large school of bluefish in my net, [a] school of these sharks attacked it from all sides and ate or liberated the school of blue-fish, practically ruining the net. Again in July, 1914, on Lookout Shoals, [I] had [a] large net filled with blue-fish attacked by [a] school of about 200 of these vicious sharks and the net ruined. [I] killed about twenty of them with harpoon and lance. (Nichols and Murphy 1916, 22)

Interactions with humans are generally of no consequence. While sandtiger sharks are found in surf zones, they do not seem to interact at all with swimmers or surfers. They are occasionally caught in the surf and, because of their size and weight, can be formidable fighters. More often they are seen by divers or caught by fishers working on wrecks or rubble piles. Spear-fishers will often encounter sandtiger sharks and are more likely to have an incident than any other group, as the sharks attempt to take speared fish from the spear. Divers reluctant to give up their catch may provoke the shark to become more aggressive, though reports of problems with sandtiger sharks are rare. The International Shark Attack File does not list any attacks from 1882 to the present that can be attributed to sandtiger sharks in Florida or the Bahamas. Only 29 unprovoked attacks have been reported worldwide, with only two fatalities.

Conservation and Management Status

Sandtiger sharks are listed as a vulnerable species by the IUCN. They are fished mainly for their meat, which is highly prized in Japan. Other fisheries along the East Coast of the United States and in Australia once reduced the shark's populations significantly. The species now enjoys widespread protection in Australian waters, and its populations are monitored closely in the US fishery to prevent population collapses.

Silky Shark

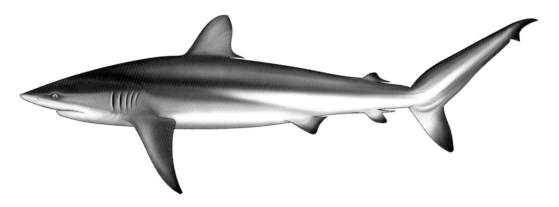

Common Name: silky shark
Scientific Name: *Carcharhinus falciformis*
Identifying Features: Silky sharks can grow to large sizes and are streamlined in appearance. They have a relatively long snout that is very slightly pointed and a head that appears to be slightly flattened. Their first dorsal fin is sloped toward the rear of the animal and the second dorsal is very small; it may actually be smaller than the anal fin. The pectoral fins are long, slender, and somewhat pointed. There may be darkly shaded fin tips on the pectoral fins, and a dorsal ridge is present between the first and second dorsal fin. Silky sharks' color is usually a dark gray to gray-brown that may also appear bronze. The underside is lighter. The skin is very smooth and is the source of their common name.
Frequency: rare

Silky sharks, along with blue sharks (*Prionace glauca*) and oceanic whitetip sharks (*Carcharhinus longimanus*), are considered to be among the more common sharks in the open oceans. Silky sharks' skin is very smooth because their dermal denticles are very small, very dense and tightly arranged, and have some overlap. The skin almost appears to have a sheen when it reflects the light in shallow water.

Range, Distribution, and Habitat Preference

Silky sharks are very common tropical and subtropical sharks found in the Atlantic, Pacific, and Indo-Pacific Oceans. They prefer open warmer waters above 73°F (23°C), though they have also been observed in shallower coastal waters. These short inshore excursions may be directed to nursery grounds when they are preparing to give birth. Their preferred habitat seems to be near continental shelves and over deep reefs or seamounts. Longline fisheries have reportedly taken silky sharks from depths as great as 13,000

The ghostly appearance of the silky shark from the open ocean shows its streamlined and well-proportioned silhouette.

Occasionally found near shore, silky sharks seem to prefer a more pelagic way of life and deeper waters surrounding islands and the continental shelf. They will frequently appear from the ocean depths to take a bait.

feet (4,000 m), indicating a truly open-ocean range. More common depths may be 600 to 1,600 feet (200–500 m).

Size and Reproduction

Silky sharks may reach a length of 9 to 10 feet (275–300 cm) and a weight of several hundred pounds. The IGFA lists an all-tackle record of 9.2 feet (279 cm) and 762 pounds (346 kg). This animal was taken off southern Australia. More commonly, silky sharks are closer to 6 feet (185 cm) in length.

Comparatively little is known of the reproductive biology of silky sharks, an open-water pelagic species. Studies have shown that males become mature at around 6 feet (180–190 cm) in length and females at a bit larger size. There has been some evidence that silky sharks may become mature at different sizes in other parts of the world, reaching lengths of 7 to 8 feet (220–245 cm) before they mature. Such lengths would correspond to animals that are 10 to 12 years old, but they may be overestimated. Some estimates place silky sharks' age at maturity at earlier ages, ranging from 6 to 10 years. Life expectancy may be at least 22 years or longer.

The silky shark is a viviparous shark, in which there is a placental attachment of embryos to the mother. Live pups are born after a gestation period of approximately 1 year. Females may give birth at 2-year intervals or longer. Litter sizes are in the range of 2 to 14 pups, whose length at birth ranges from 23 to 30 inches (60–75 cm), though embryos of larger sizes have also been observed.

Food and Feeding

With the silky shark's preference for open waters and its pelagic life style, one would expect that it would prey on animals with similar patterns. Tuna, mackerel, albacore, sardines, and reef fish such as groupers and mullet have been found in stomach contents. The silky shark also will consume squid. One study analyzed 786 silky shark stomachs as part of a tuna purse seine fishery and found that skipjacks and yellowfin tuna made up more than 50% of the food the sharks had consumed. Flying fish, jacks, and halfbeaks were also consumed in large numbers, as well as squid. This study concluded that juvenile and adult silky sharks eat basically the same type of fish.

Behavior and Interactions with Humans

Silky sharks often travel in schools, with sharks of similar sizes or with other silky sharks of the same sex. Silky sharks have also

An extremely fast open-water shark, the silky shark is fast and agile and capable of consuming tuna and high-speed predatory fish.

Silky sharks have been heavily fished by the commercial industry. Purse seiners, probably targeting tuna, often find in their nets silky shark that are probably also targeting schools of tuna. Silky sharks are also taken by longline fishers and occasionally by recreational fishers. They are surprisingly tolerant of injuries, and as long as their ability to feed effectively isn't compromised, even fish with large hooks in their jaws can probably survive until the hook rusts away. Many sport fishers often use barbless hooks made of highly corrosive steel alloys that easily rust and degrade instead of the longer-lasting stainless steel hooks.

been seen giving threat displays to divers and swimmers. The displays include arching the back and pointing the pectoral fins and often are accompanied by some general shaking of the body. The same displays have been seen in other species and are known to be predictive of a territorial dispute or a threat. The best course of action is to back away from sharks showing these behaviors, to prevent the display from becoming something more aggressive.

The silky sharks' preferences for open water make contact with humans of little concern. While they may venture near shore in Florida and the Bahamas when they give birth, their residence time may be short and insufficient for human encounters. There have been no recorded attacks in Florida or the Bahamas. But since silky sharks can grow large, they should be shown the same level of respect due to any large shark.

Conservation and Management Status

The worldwide fishery for silky sharks has been intensive. Though the species is plentiful, its presence as bycatch in purse seines has raised some concern. One study showed that silky shark mortality from capture by purse seine fisheries was nearly 81%. More innovative nets might address and eventually resolve this issue. The sharks are not wasted, though. Their fins are regarded as valuable, and they are sold in addition to the tuna that are targeted.

In spite of this fishery, worldwide numbers are thought to be stable. The silky shark is thus considered "near threatened" by the IUCN and is not in need of more stringent controls or management practices throughout most of its range. The northwestern and western central Atlantic populations have shown decreasing numbers, and these populations are classified as "vulnerable."

Smoothhound Sharks

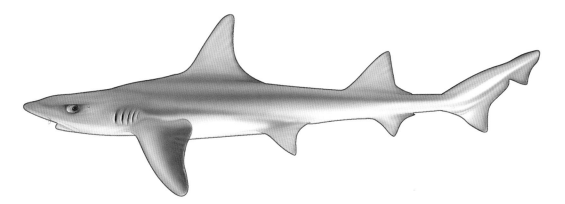

Common Names: smooth dogfish, dusky smoothhound; Florida smoothhound or Florida dogfish; gulf smoothhound or Gulf of Mexico smoothhound

Scientific Names: *Mustelus canis* (smooth dogfish or dusky smoothhound); *Mustelus norrisi* (narrowfin or Florida smoothhound); *Mustelus sinusmexicanus* (Gulf or Gulf of Mexico smoothhound [rare])

Identifying Features: Smoothhound sharks are small and slender. They range in color from a dark greenish or olive-gray to brown with a lighter underside. The dorsal fins are almost the same size, though the first dorsal is slightly larger than the second. The shape of the lower lobe of the caudal fin (the tail) varies between species. Their snout is tapered and slightly pointed. They also have a spiracle just behind the eyes. Their head is notably flattened compared to other sharks' heads. There is also a dorsal ridge between the two dorsal fins. Distinguishing between the species is difficult unless they are side by side.

Frequency: common, except for the Gulf smoothhound shark

A relative of these small elusive sharks—the spiny dogfish shark (*Squalus acanthias*)—may be more familiar to many people than the smoothhound sharks themselves. Though not commonly found in Florida or Bahamian waters, the spiny dogfish is the species many students dissect in high school or college biology classes. The smoothhound sharks are distant cousins that lack the spines but are generally the same size and show the very familiar body shape of the more common dogfish shark. They are very common in the shallow inshore waters of Florida and the Bahamas. Very little is known about the Gulf of Mexico smoothhound shark, other than that it can be found in the Gulf of Mexico overlapping with other smoothhound shark species.

The slim and sleek profile of the dusky smoothhound shark resembles its distant cousin, the spiny dogfish shark, which is commonly studied in comparative anatomy classes. Its disproportionately large eyes, equal-sized first and second dorsal fins, and notched dorsal fin are keys to its identification.

Range, Distribution, and Habitat Preference

The dusky smoothhound shark (*Mustelus canis*) is widely distributed along the East Coast of the United States and is also found in the Bahamas and the Gulf of Mexico. It prefers shallow inshore waters, though the animals have been taken from depths up to 650 feet (200 m). The narrowfin smoothhound shark (*Mustelus norrisi*) is not as widely distributed but has been caught along the Florida east coast and in the Gulf. It may migrate more than the dusky smoothhound shark and therefore may not be encountered as frequently, though it, too, prefers shallow inshore waters. Scientists are not completely certain that the dusky smoothhound shark and the narrowfin smoothhound shark are actually separate species; geneticists are working to resolve this question through analysis of DNA from these species.

The gulf smoothhound shark (*Mustelus sinusmexicanus*) has only recently been described. Very little is known of its distribution or its biology. It is known to be mostly confined to the Gulf of Mexico and to prefer waters near the continental shelf and slope in water deeper than either of the other two species inhabit. As such, it is likely to be encountered only by fishers who are bottom-fishing in deeper waters or around deep-water oil rigs; both fishers and divers may find the sharks there.

Size, Age, Growth, and Reproduction

Dusky smoothhound sharks have been known to reach almost 5 feet (150 cm) in length, though most animals average closer to 4 feet (125 cm). The maximum size for the narrowfin shark is about 43 inches (110 cm). More commonly, the animals are 30 to 35 inches (75 to 90 cm) long.

The growth rates are relatively rapid, between 17 and 27 inches (40–70 cm) in 2 to 4 years. The sharks are estimated to live from 10 years for males to 16 years for females.

Males reach reproductive maturity at 2 to 3 years and lengths of 31 to 35 inches (80–90 cm), and females at 4 to 7 years and a length of 35 inches (about 90 cm). The young are born live and are nourished by a yolk sac placenta. Gestation lasts for about 10 months, and litter sizes range from 4 to 20 pups. The young are born at lengths of 11 to 15 inches (30–40 cm) for dusky smoothhound sharks

Gulf smoothhound sharks are typically found offshore in deep waters, often near the edges of the continental shelf. They are frequent visitors to offshore oil rigs.

and about 12 inches (30 cm) for narrowfin smoothhound sharks. The narrowfin smoothhound shark litter size may also be somewhat smaller, ranging from 7 to 14 pups.

Food and Feeding

The teeth of smoothhound sharks are small, flat, and blunt, well adapted to crushing their prey. This type of dentition may explain the shark's preference for crabs, lobsters, shrimps, and other hard-bodied prey items. Small fish are also a part of their diet.

Since they seldom grow to large sizes, they are a common food item for other species of sharks. They have been found in the stomachs of dusky sharks, hammerhead sharks, great white sharks, sandbar sharks, and blacktip sharks.

Behavior and Interactions with Humans

Smoothhound sharks show some schooling behaviors and some short-distance migration. Populations in the northeastern United States and in the Gulf are thought to be isolated from each other because they do not come into contact through migration and thus these populations show very little tendency to intermix.

Interactions with humans are rare because of the shy nature of smoothhound sharks. It is likely that any encounters with them that result in human injuries come from improper handling and release of animals taken by fishers.

Conservation and Management Status

Smoothhound sharks are not considered to be facing any particular threats from fishing pressures. They are taken as bycatch and seldom by a directed fishery. Because so little information is available for smoothhound sharks, dusky smoothhound sharks are listed as "threatened" and narrowfin smoothhound sharks are listed as "data deficient" by the IUCN.

Spinner Shark

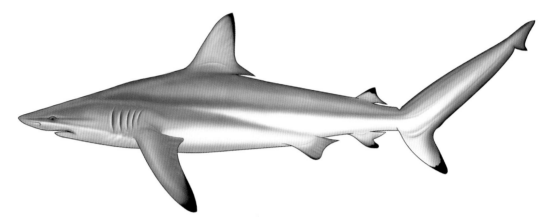

Common Names: spinner shark, large black-tip

Scientific Name: *Carcharhinus brevipinna*

Identifying Features: Spinner sharks are slender and have a long, pointed snout. They lack a ridge between the first and second dorsal fins. Their coloration ranges from brownish to brown-gray with a white underside, and the tips of the fins are black. The black coloration of the anal fin distinguishes the spinner shark from the blacktip shark, whose anal fin does not have a black tip.

Frequency: common

Spinner sharks are elusive and shy, so they are not photographed as easily as other species. They are frequently misidentified as blacktip sharks. The anal fin of spinner sharks is tipped in black, as shown in this photo, while the anal fin of the blacktip shark is white.

Spinner sharks are extremely fast-swimming and are prized by game fishers for their acrobatics when hooked. They often leap clear of the water, and their flips and spins explain their common name. There are stories of these sharks showing the same leaping and spinning behaviors when they feed on schools of fish.

Range, Distribution, and Habitat Preference

Spinner sharks are found worldwide They prefer shallow inshore waters in temperate and tropical zones, though specimens have reportedly been taken from depths of 240 to 300 feet (75–100 m). Spinner sharks are very common in all Florida waters and in the Bahamas. Because of its close resemblance to the blacktip shark, the spinner shark's actual distribution is not well understood. It is known to prefer nearshore waters, and this preference permits easy observation of its long-distance migrations, which occur in very large schools.

Size, Age, Growth, and Reproduction

The maximum size of spinner sharks is around 8 to 10 feet (240–300 cm), though the larger animals are less common. They are

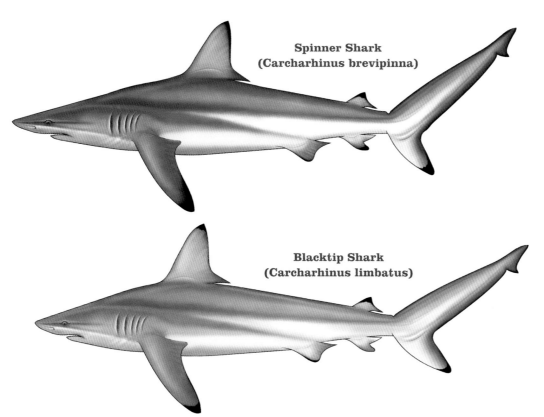

**Spinner Shark
(Carcharhinus brevipinna)**

**Blacktip Shark
(Carcharhinus limbatus)**

A side-by-side comparison highlights the differences between spinner sharks (*top*) and blacktip sharks (*bottom*) in this illustration. It is easy to understand why identification of the animals is often confused.

more frequently encountered at lengths of about 6 feet (180–185 cm). Males are thought to become sexually mature at around 5.5 feet (170 cm). Females are slightly larger, maturing at 6.5 feet (195 cm). Estimates from other parts of the world are that spinner sharks become a bit larger before they mature, with mature males and females both reaching slightly more than 7 feet (220 cm). No explanation has been given for these differences.

Juvenile spinner sharks grow at a remarkable rate during their first year. They may increase in length by 10 to 12 inches (25–35 cm) in their first 6 months and may even double their length within the first year. Thereafter the rate drops steadily to an annual change in length of 2 to 4 inches (5–10 cm) per year. They are thought to live at least 11 years but may survive up to 20 years or more.

Male spinner sharks reproduce when they are 4 to 7 years old, and females when they are 7 to 8 years old. Once they mature, female spinner sharks are thought to reproduce on a 2-year cycle. The gestation period has been reported to be 11 to 15 months. The average litter size is 6 to 10 pups, who are nourished through a yolk sac placenta and are born live at a length of 22 to 25 inches (55–65 cm).

Food and Feeding

The favorite food of spinner sharks is fish of various species. Their migrations may actually be synchronized with the movements of some of their preferred food, such as bluefish, mackerel, herring, menhaden, and sardines. They also consume grunts, jacks, stingrays, and squid. The IGFA world record was taken with bluefish as bait. However, stomach-content studies have revealed a wide variety of other food items, including squid, octopuses, and other cephalopod mollusks.

Behavior and Interactions with Humans

The schooling behavior of spinner sharks and their often visible migrations are their most notable behaviors, in addition to their acrobatic displays. Feeding in these large schools has reportedly often resulted in mass feeding frenzies, particularly around shrimp boats, which they frequently follow to seize the bycatch that is cast overboard.

There have been 15 recorded unprovoked attacks on humans directly attributed to spinner sharks, according to the International Shark Attack File. Because of their proximity to shore during their migrations, they are likely to be found in the surf zones and may mistakenly reach out to a swimmer or surfer. Similarly, divers spearing fish in waters where spinner sharks are found may lose their speared fish to a spinner shark that is passing by. Here again, misidentification may attribute an attack to a spinner shark when it was actually a blacktip shark that attacked. Actual identification is usually more important to a biologist than to a victim of a shark bite.

Conservation and Management Status

The spinner shark is not well studied in the commercial fisheries of the world, because it is so often confused with the blacktip shark. The market preference for meat from blacktip sharks and the value of their fins might mean that spinner sharks, even if properly identified, are sometimes sold as blacktip sharks just to gain entry to the marketplace.

Recreational fisheries for spinner sharks are also thought to be significant. But in recreational fisheries, just as in commercial fisheries, misidentification of spinner sharks as blacktip sharks may mean that the actual number of spinner sharks that are landed are underreported. Nevertheless, the IUCN lists the spinner shark as "near threatened" and does not suggest any particular recommendations for population management.

Tiger Shark

Common Name: tiger shark

Scientific Name: *Galeocerdo cuvier*

Identifying Features: Tiger sharks are relatively easy to identify because of their unique color patterns. The stripes of tiger sharks are retained as they grow, though they may fade as the animal reaches its maximum sizes. The shark's blunt snout and the lateral keels near its tail are also key characteristics. Perhaps its best identifying feature is the shape of its teeth—not especially useful information for divers who may be confronted by such a large animal. The teeth are deeply notched and angular, very different from the long, pointy or triangular-shaped teeth of many other shark species.

Frequency: very common

License plates, baseballs, cows, goats, turtles, birds, and even a suit of armor have been found in the stomachs of tiger sharks. The suit of armor may be a bit of an exaggeration. But if something strange is reported in a shark's stomach, there is a good chance it was from a tiger shark. Tiger sharks have been called indiscriminate feeders because their diets are so varied, but opportunistic is probably a more fitting description of their willingness to feed on whatever might be available.

Range, Distribution, and Habitat Preference

Tiger sharks are found worldwide in temperate and tropical seas. They are very common throughout Florida and the Bahamas and are likely to be seen in deep waters well offshore to shallow coastal areas, including bays and estuaries. Satellite tracking shows that tiger sharks are capable of very-long-distance migrations. Some of these studies have revealed that even after long migrations, tiger sharks may return to the same area. What drives the migration and what factors are responsible

The clear water and white sandy bottom serve to accent the stripes that characterize the tiger shark. The stripes may fade in adulthood, but remnants are visible even in the largest specimens.

A juvenile tiger shark cruises near the surface in the shallow waters of the Bahamas. Its underside is much lighter than its surface and almost masks its appearance from underneath. The darker surface coloration helps it to blend in better with the bottom when viewed from above. This darker upper surface and lighter ventral surface, called counter-shading, is a common disguise in fish.

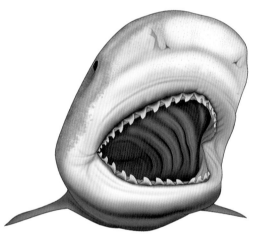

The massive size of the tiger shark's jaw is shown in this illustration. The heavily angled teeth are well equipped to easily tear flesh from the animals that make up its wide and diverse diet.

for the sharks' return to the same area is not understood.

Size, Age, Growth, and Reproduction

Tiger sharks are among the largest sharks. Only three or four other species of sharks reach the lengths reported for a full-grown tiger shark. Accurate estimates put the maximum size at 13 to 15 feet (4–5 m), though some estimates have been made of lengths greater than 15 feet (5.5 m). A tiger shark of 13 feet (4 m) in length would be massive and might weigh as much as 1,300 pounds (600 kg). The IGFA lists the all-tackle world record at 1,780 pounds (807 kg), with a length of 13.9 feet (423 cm), caught off of South Carolina.

Tiger shark pups are born alive, though, unlike other closely related species, there is no placental connection to the mother during the development of the young. Instead, the young hatch internally, are presumably nourished by secretions from the mother's uterus, and then are born live when embryonic development is complete. It is estimated that gestation may last up to a year and that litters may range from 10 to 82 pups. The reported sizes of newborn tiger sharks range from 20 to 35 inches (50–90 cm), but most of the records support the larger sizes. Reproduction is thought to occur every 2 to 3 years.

Tiger sharks are thought to reach reproductive maturity at a length of 8 to 9 feet (300 cm) and an age of 7 to 8 years. Life expectancy may be between 27 and 37 years.

Food and Feeding

The diet of a tiger shark is as diverse as that of any other shark species. Tiger sharks will feed on fish of almost any species, including bony fish, sharks, and rays. They also pursue dugongs, birds, sea turtles, squid, lobsters, conchs, and other mollusks. It seems that their diet has adapted to feed on whatever animals happen to be in abundance wherever the sharks may be. Drowned terrestrial animals such as cows and horses have also been reported from studies of tiger shark stomach contents, along with assorted cans, bottles, and plastics that apparently were mistaken by the shark as food items. Unfortunately, human remains have also been discovered in their stomachs.

Interactions with Humans

Tiger sharks are the best-known inhabitants of Tiger Beach, a popular dive site in the Bahamas. Drawn to the area by baits deployed by dive boats, large tiger sharks approach divers even to the point where physical contact occurs in nonthreatening encounters. Divers are instructed prior to the dives to follow procedures that aim to minimize the possibility of attacks or injuries. These dives have become very popular with underwater photographers, and some of the images of interactions with tiger sharks are breathtaking.

Not all encounters with humans are so benign. Tiger sharks have been implicated in many attacks worldwide and are probably best known for fatal encounters in Hawaii. These widely publicized attacks, many of which were fatal, prompted the Hawaiian government to undertake a large-scale fishing effort to catch and kill these large sharks. The control efforts encountered difficulty when it was revealed that killing the tiger sharks conflicted with the cultural heritage and beliefs of native Hawaiians. In Florida, five confirmed, unprovoked attacks by tiger sharks have been reported to the International Shark Attack

The massive size of the tiger shark ranks it as one of the largest species of sharks. Fishing records include a nearly 14-foot (4.3 m) tiger shark with a weight of just under 1,800 pounds (636 kg), which is an exceedingly large weight for an animal of that length.

Green turtles as well as other common marine turtle species are frequently found with serious injuries believed to result from encounters with tiger sharks. Turtle fragments are often discovered in the stomachs of tiger sharks. The sharks' powerful jaws and heavy teeth equip them to crush and tear through the protective shell that shields turtles from most other predators.

File. Two of these attacks were fatal. Only one unprovoked attack has been recorded from the Bahamas, and it resulted in a fatality.

Conservation and Management Status

Tiger sharks are listed as "near threatened" by the IUCN. Most populations are not experiencing undue fishing pressures, and worldwide stocks appear to be stable. Tiger shark fins, skin, and livers are valuable, and some commercial fisheries target them for these products. In some regions, where fishing pressures are greater or where bycatch may be higher, efforts have been undertaken to ensure that local populations remain stable.

The IUCN has expressed concern that the sometimes indiscriminate feeding of tiger sharks on garbage and other human debris might become more of a threat in the future and may surpass threats from overfishing.

Tiger sharks are seemingly fearless and are not timid around divers. Many shark encounters center on attracting tiger sharks because of their charismatic reputation and their photogenic image.

In spite of the tiger shark's previous reputation as a dangerous shark to be avoided at all costs, a growing ecotourism market has exposed thousands of divers to these sharks. Under rigidly controlled conditions, encounters with tiger sharks and hammerhead sharks have been remarkably free of incidents and have provided close-up opportunities for sport divers and photographers alike.

A tiger shark with its entourage of remoras cruises the shallow inshore grass flats, also a favorite haunt of marine turtles. In other parts of the world, such encounters form the basis of several unusual food chains.

There is little doubt that the tiger shark reigns as one of the dominant shark species in the shallow waters of Florida and the Bahamas.

The graceful silhouette and bulk of the tiger shark are easily apparent from a ventral view that also showcases its oversized mouth.

Whale Shark

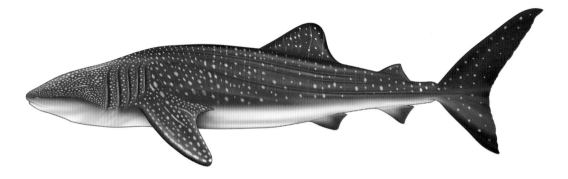

Common Name: whale shark

Scientific Name: *Rhincodon typus*

Identifying Features: Whale sharks are very large and have a blunt, flattened head. They are gray to brownish in color with yellow or whitish spots. A whale shark's mouth is located at the front of the snout, and there are very prominent ridges running along the sides. Its size, as the largest shark species, and its white-spotted skin make it easy to identify. In fact, the pattern of spots is different for each individual shark. Identification is often made by photographing the animal and comparing its pattern to a database of known animals. This tool is especially important as an aid to identify whale sharks that return to the same location periodically. If they can be dependably identified, elements of their life history patterns can be more easily determined.

Frequency: rare

As the largest fish in the sea, the whale shark dwarfs this diver, who is attempting to fit the entire length of the shark into a single photograph. Whale sharks are known to exceed 40 feet (12 m) in length.

"You're gonna need a bigger boat." These words from the movie *Jaws* were spoken in reference to a great white shark. But they are even more appropriate to describe whale sharks, which hold the distinction of the largest fish in the sea. The impact on movie-goers, however, would not seem as startling if the hunters were seeking a large plankton-feeding shark rather than a supposed rogue, man-eating species.

Range, Distribution, and Habitat Preference

Whale sharks are found worldwide, preferring tropical and near-tropical waters, where water temperatures are reliably above 70°F (21°C). Large aggregations have been found in the Philippines and around Mexico, South Africa, and Australia. They are not very abundant in Florida and Bahamian waters. When they are sighted, it is generally offshore in deeper water, though some have been seen inside the reefs in the Florida Keys and some have been discovered beached in these same

islands. Reports of whale sharks around An-
dros Island and the Exumas in the Bahamas
indicate that they will venture into shallow
waters on occasion. Fishers who prefer to
troll for large game fish at longer distance
offshore are more likely to spot them than are
beachgoers or divers.

Size, Age, Growth, and Reproduction

Lengths in excess of 40 feet (12.2 meters)
have been measured, though there are reports of larger animals. Exact measurements
of animals reported to be larger are absent,
and these are generally regarded as estimates.

Most studies of growth rely on measurements taken during initial tagging and later
recapture and remeasurement. The difficulty
of obtaining an accurate length for such a
large animal makes estimates of growth very
difficult, if not impossible. The few estimates
of growth that exist come from a few captive

studies and show an annual growth of 8 to 11
inches (21.6–29.5 cm) per year. Mathematical
models predict that whale sharks could live
anywhere from 60 to 100 years, though these
life spans have not been validated.

Computerized scans of the patterns of
spots on whale sharks indicate that individual sharks can be recognized from their
unique pattern. The eventual creation of a
worldwide database indexing the spot pattern for every whale shark encountered may
make tracking and life history studies more
revealing. Other advances in technology may
make underwater measurements of total
length possible. The combination of these
techniques may ultimately allow for the determination of accurate growth rates from
these data. Better insights into life history,
gained from wild populations rather than
captive sharks, may then become possible.

What sparse information is available re-

The unusual perspective
of a wide-angle lens
distorts the true size
and exaggerates the
proportions of this
filter-feeding giant, but
it does shows the mam-
moth volume of water
that can be filtered of its
plankton by a feeding
whale shark.

Whale sharks are drawn to certain locations for a variety of reasons. Their migrations often arrive at specific locations in synchrony with massive plankton blooms. There is also speculation that their migratory behavior, like that of other shark species, may coincide with mating or with giving birth. Research projects are actively attempting to identify whale sharks' mating and nursery grounds. *Photo courtesy Oscar Reyes. Used with permission.*

garding reproduction in whale sharks comes from studies of pregnant animals captured when commercial fisheries have landed these animals. In what might rank as the most important study of this type, the dissection of a pregnant female revealed 300 young in varying stages of development. The study also provided more support of aplacental viviparity in whale sharks. This is the same developmental strategy used by nurse sharks, a close relative, in which the eggs hatch inside the female when the embryos have developed sufficiently. The egg cases are then shed, and the live young are born sometime later. A litter of this size is the largest reported for any shark species.

Whale sharks are reported to reach sexual maturity at lengths between 29 and 32 feet (9–10 m) and are born at a length of about 2 feet (60 cm).

Food and Feeding

Whale sharks feed primarily on plankton, the ocean's smallest creatures, by filtering water through their gills to extract the plank-

ton from the water. This is the same feeding mechanism used by large whales. Whale sharks can be seen swimming near the surface, where the highest density of plankton is found. Recent studies have revealed that whale sharks will also make foraging dives to as deep as 1,550 to 1,600 feet (475–500 m), where it is thought that they feed on deep-water animals as well.

Unlike whales and some other plankton-feeding sharks, whale sharks are capable of actively sucking in large volumes of water that may contain larger aquatic organisms. Suction feeding is also a characteristic of nurse sharks, one of the whale shark's closest relatives, though the food items differ considerably. Whale sharks have been observed near the surface in a vertical position with their mouths near the surface while they slurp in huge volumes of water. This feeding behavior may explain the presence of small fish, squid, fish eggs, small crustaceans, and occasional masses of algae in additional to the more common planktonic organisms that have been reported from studies of their gut contents.

Whale sharks are known to undergo long-distance migrations. Tagging programs and satellite tracking studies are beginning to unravel the movement and migratory patterns of these large sharks. How they are able to arrive at locations at the precise times when massive plankton blooms occur is as vexing

This would be the last image seen by a small fish or planktonic organism that happens to be in front of one of these seagoing strainers.

for the study of whale shark behavior as it is for large marine mammals.

In spite of whale sharks' large size, some aspects of their lives are unknown. Aggregations of hundreds of individuals are commonly encountered, but the reasons for such aggregations are not widely understood. Feeding during plankton blooms of various types is generally assumed to explain why these sharks are found in such large numbers. Although other species of sharks are known to gather in large schools for mating or near nursery grounds where they give birth, no evidence of such purposes exists for whale sharks. Nursery grounds and mating grounds are yet to be discovered. The sharks' large size makes capture nearly impossible, so studies of captive animals are very rare. Few facilities have holding tanks large enough to provide a suitable space for such massive fish. Whether behaviors in captive animals would be identical to those of free-living animals is also a concern. Mating and pupping by large shark species in captivity is also extremely rare.

Interactions with Humans

A very large ecotourism industry has developed for people who want to swim with whale sharks. They have not been shown to be dangerous to humans, other than from their large size, and swimming with them has become very popular. In some parts of the world where whale sharks were once taken commercially, the ecotourism industry has provided so much income that governments have begun to realize that protecting the animals is more profitable financially than taking the animals for food or other commercial purposes. In those locations, protective measures have been enacted to prevent damage to the expanding tourist trade.

Despite their size, some whale sharks have been captured for large aquariums, though few captive facilities have tanks that can hold animals that reach 20 to 30 feet (6–9 m) long. Exhibits that can contain these fish attract a

Whale sharks differ from many other true filter feeders in that they also utilize suction feeding to gulp in huge quantities of plankton-laden water. The filtering mechanism is just as effective in removing plankton from the surface waters.

Graceful giants of the sea, whale sharks are seeing greater levels of worldwide protection.

large number of visitors. Visitors often participate in feeding and are able to witness the gulping and filter feeding that occurs as the sharks consume the large volumes of krill that are offered during feeding times.

Conservation and Management Status

Once the target of fishery pressures, whale sharks are less important to commercial fisheries than previously, partly because of the decline in their numbers. Their value to ecotourism in many regions exceeds their value to local fisheries, and many countries have implemented protective measures to ensure population stability. The IUCN lists the whale shark as "vulnerable," one of the categories of animals that are considered to be under threat, partly as a result of their decreasing numbers and poor reproductive strategy.

PART THREE

Skates and Rays, Including Sawfish

Sharks, skates, rays, and sawfish share many biological characteristics and are distantly related. Because they also occupy many of the same habitats in Florida and the Bahamas, fishers, divers, boaters, and beachgoers who look for sharks are very likely to encounter skates, rays, and occasionally sawfish, a close cousin to the rays. For that reason, a brief series of accounts is provided to satisfy the curiosity of readers who may encounter these aquatic creatures and to aid in understanding the similarities as well as the differences between the sharks and these relatives of theirs.

Atlantic Stingray and Southern Stingray

Common Names: Atlantic stingray, southern stingray

Scientific Names: Atlantic stingray: *Dasyatis sabina;* southern stingray: *Dasyatis americana*

Identifying Features: Most rays, including the Atlantic stingray, have bodies that are rounded or disk-shaped. The southern stingray is an exception to this general body form. Southern stingrays' bodies are more diamond-shaped, and the tips of the pectoral fins are more pointed. In both species, the body rises from the wings, and the animals have prominent eyes and spiracles on the surface.

The spiracles are a special adaptation that allows the ray to bring a current of oxygen-containing water into the gill chamber to breathe. Their location on the upper body surface means that the animals can remain mostly buried under the sand and still breathe even though their gills, located underneath the animal's body, are buried in the sand.

Both species are brown to brown-yellow, even an olive-green on the upper surface, and lighter-colored on the underside. The tail of the Atlantic stingray is not as long as the southern stingray's. It may reach a length as long as the Atlantic stingray's body, while the southern stingray's tail may grow to about twice the length of its body. Tails do not always remain intact, so using tail length as a guide to identification is not a way to positively identify the rays.

Frequency: Atlantic stingray: very common in Florida, absent from Bahamas; southern stingray: very common

The Atlantic stingray is a small species of ray found near shore and very commonly in brackish and freshwater lagoons.

The southern stingray is a larger species than the Atlantic ray. It is extremely common along the shore and around reef systems. The body is less rounded and more diamond-shaped than that of Atlantic rays.

Southern stingrays are a common sight in shallow clear waters. Numerous tourist enterprises in southern Florida and the Bahamas cater to snorkelers and novice divers' desires to experience these animals in close-up encounters. Schools of rays often surround the swimmers and seem to clamor for attention. These activities are benign, with injuries essentially nonexistent. In fact, the rays are so accustomed to divers that they will often disrupt other diving activities.

Atlantic and southern stingrays are very closely related, and most of the information about their biology is similar, if not identical. They are treated together here, and where differences are known, they are highlighted.

Stingrays are bottom-dwellers. They are well adapted for life in the sands and muds of the tropical and subtropical waters of Florida and the Bahamas. Their spiracles, openings to the respiratory system placed on the head just behind the eyes, allow them to breathe even when their gills, located on their underside, are buried in the sand. Their large pectoral fins, or wings, may help them surround and trap prey items that can then be captured and consumed, using their crushing jaws. Their large wings are used for propulsion, their undulating, wavelike movements propelling the animals through the water at sometimes remarkable speeds.

Range, Distribution, and Habitat Preference

Southern and Atlantic rays are tropical and subtropical species that are very common in Florida and the Gulf of Mexico. Southern rays extend to the Bahamas, but Atlantic rays are not found there. Both species are most commonly encountered in shallow, nearshore waters, generally preferring water that is less than 165 feet (53 m) in depth. They are common over inshore sand and grass areas as well as around coral reefs. Atlantic rays are well known as inhabitants of freshwaters around Florida. They are commonly found in the Indian River Lagoon system and penetrate far up the St. Johns River in northern Florida. The close proximity to shore of both species often results in accidents to beachgoers and waders who may step on partially buried stingrays and receive a nasty wound from the barb at the base of the tail.

Southern rays commonly surround prey items such as this queen conch and use their strong jaws and crushing teeth to pulverize the shell. Once the shell is destroyed, the conch can be sucked from the shell remains to serve as food for the ray.

Size, Age, Growth, and Reproduction

Atlantic stingrays are among the smaller ray species. Their disk width is seldom more than 12 to 13 inches (25–35 cm); they are much smaller than southern stingrays. The confusion in identity may exist only between Atlantic stingrays and juvenile southern stingrays. Males reach sexual maturity at around 8 inches (20 cm), and females at 9 inches (24 cm).

Southern stingrays can attain a size of nearly 80 inches (200 cm) wide, with weights greater than 200 pounds (90 kg). Such sizes are uncommon; disc widths less than 40 inches (100 cm) are more common. Little is known of their age and growth. Captive studies have shown that females reproduce when they reach an age of around 5 to 6 years and a disk width of 30 to 32 inches (75–80 cm). Males reach maturity at 3 to 4 years old and a disk width of 20 inches (50–55 cm). Their life expectancy is unknown.

Mating stingrays have been seen in the wild but have not been systematically studied. Some captive mating has been observed, and the information that is available comes from these studies. The fertilized stingray eggs are known to develop by aplacental viviparity, according to which the embryos hatch inside the mother, where they are nourished

by a yolk sac and from secretions produced from the uterine lining of the mother. These secretions are known as histotroph or by the more common name of uterine milk. The young are born live after a gestation period that is estimated to average 175 days in southern stingrays and about 4 months for Atlantic stingrays. The litters may range from one to four in Atlantic stingrays, and their length at birth is about 4 to 5 inches (10–13 cm). Southern stingrays have somewhat larger litters, averaging 2 to 10 young stingrays that are about 6 to 10 inches (17–25 cm) long at birth. How frequently they mate and give birth is also unknown, though some studies indicate

The southern ray is more easily discovered when sighted over a grass and rock bottom. There it is not as well camouflaged as when it rests on a mud or sand bottom.

they may reproduce twice a year in some locations.

Food and Feeding

Southern stingrays are generally classified as foragers, constantly wandering through their habitat in search of suitable prey. Their stomach contents are varied. Many types of crustaceans, such as crabs, shrimps, and lobsters, are commonly found when their stomachs are examined. Mollusks such as clams and conchs have been found, and small fish are common. Their diet is diverse, and they are often considered to be opportunistic feeders, consuming whatever might be readily available. Their habit of fanning the bottom sands with their large wings to expose buried prey provides them with an array of bottom-dwelling organisms that might be overlooked by other predators.

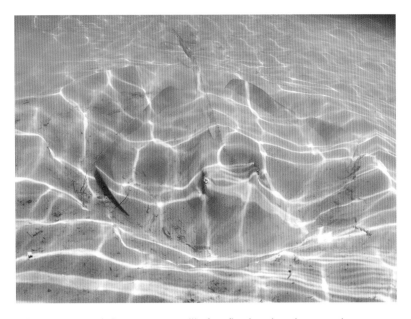

When over a sandy bottom, a ray will often flap its wings in a way that disturbs the sand and covers the ray's back, making the ray undetectable to predators relying only on vision to locate it.

Stingrays are a favorite food for hammerhead sharks, which have often been seen engaged in high-speed chases after large stingrays. Eventually, the shark pins the stingray to the bottom and uses its large head to hold the stingray in place while it bites off large chunks of the stingray's wings. Even when buried under the sand, the stingrays can be detected by passing hammerhead sharks, who seem to take particular delight in consuming such a large meal.

Interactions with Humans

Stingrays are docile creatures and are often wary of humans. Encounters are common and are usually concluded by the stingray, which flees amid a huge plume of sand. Smaller stingrays are often near the shore, and because their habit is to bury themselves under the sand, they are frequently stepped on by wading beachgoers. If the unsuspecting wader should step on the barb, a nasty puncture wound can result. The barbs are sharp and may cause severe puncture wounds and a large amount of bleeding. It has been estimated that as many as 1,500 stingray injuries per year occur in the United States. Injuries are rarely fatal unless the barb penetrates the chest cavity, as it did with television star Steve Irwin. If the wound is not treated, the venom may have delayed effects, including tissue destruction.

The American College of Emergency Physicians notes that the pain caused by the spine and the venom is uncharacteristically

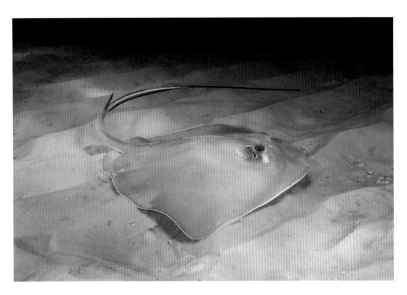

The lighter coloration blends well over sand and offers another level of protection from predators such as hammerhead sharks, which seem to actively seek and pursue even the largest stingrays. The barb at the base of the tail is visible in this photograph.

severe for the actual extent of the injury and is attributed to the venom associated with the barb. The barbs generally have teeth on their edges that may anchor the barb in the skin and require surgical removal. The venom is known to include proteins that can be destroyed by high temperatures, so hot water is therefore thought to be effective in destroying the venom; of course it should not be so hot that it causes burns. Secondary infections from the severe nature of the puncture wound itself and bacteria that may reside on the skin may further complicate the healing process. Barbs that a stingray loses or periodically sheds may regrow.

In some areas, particularly around the Cayman Islands, stingrays are found in huge aggregations and have become a tourist attraction. Stingray City is a popular destination where aggregations of stingrays occur and are fed by hand. Tourists flock to this popular dive spot to pose, often with hundreds of large southern stingrays.

Conservation and Management Status

There is not much of a fishery for Atlantic or southern stingrays. They are mostly taken as bycatch in trawls and nets by commercial fishers and are generally released. Recreational fishers also encounter these animals, and if they are landed, they are released. The huge sizes of southern stingrays make them all but impossible to land when hooked on most types of light tackle, and they are not highly sought. Nevertheless, the IGFA actually lists a world record of 246 pounds (111.6 kg) for a southern stingray caught off Texas by a young woman in 1998. The animal was 62 inches (157 cm) wide. The IGFA all-tackle record for an Atlantic stingray is a paltry 10.75 pounds (4.9 kg), hardly comparable to the record for its monstrous cousin.

Because there is no significant fishery, most populations of southern stingrays are healthy throughout their range, and no special management measures have been enacted

The rays' preference for shallow waters near shore often places them in areas preferred by waders, swimmers, and other beachgoers. It is in these areas where a vast percentage of injuries occur, when people accidentally step on a ray and sustain a painful and potentially serious puncture wound from the spine at the base of the tail.

The southern stingrays' dark dorsal coloration makes the them highly visible when they are cruising over shallow sandy bottoms. They are often vulnerable to predation when they lack the protection of bottom sands, where they bury themselves as a disguise.

or recommended. Most stingray species have not been studied to the same extent as their shark cousins. The IUCN lists the southern ray as "data deficient," an indication that more information is needed. Atlantic stingrays are classified as species of "least concern."

Clearnose Skate

Common Name: clearnose skate
Scientific Name: *Raja eglanteria*
Identifying Features: As its name suggests, this skate has an almost transparent region on its snout, on either side of the midline. The snout also has what appears from the top to be a knob on the end. The skate's back has rows of spiny thorns that extend to its tail. The tail has two smallish dorsal fins near its tip. The color and patterns on the upper body surface range from dark brown to grayish-brown and may serve to camouflage the animal when it rests on the bottom. The underside is very light compared to the upper surface.
Frequency: common in Florida and the Gulf of Mexico; not reported in the Bahamas

The clearnose skate is named for the almost transparent portions of its snout. It is a common skate in Florida near-shore waters but is not reported to be present in the Bahamas.

Always a curiosity, the mermaid's purse is a beachgoer's treasure. Often found in large numbers on sandy beaches, these curios are empty skate egg cases, and in Florida, they are the empty egg cases of the clear-nose skate. The clearnose skate is the only species of skate that is common to Florida waters in locations where it is likely to be seen by beachgoers, waders, fishers, boater, and divers. The clearnose skates' preference for coastal, nearshore habitats explains why their egg cases are so abundant during the time of the year when they give birth. Other species are present, but they inhabit such deep waters that they are generally not encountered.

The skate has little commercial or recreational value. It is seldom taken for food, and its tiny fins have no value for the manufacture of soup. Its value to education and research, however, is tremendous. The species does well in captivity, even reproducing successfully. Zoos and aquariums rely on clearnose skates to introduce their patrons to these close relatives of sharks. Clearnose skates have also been the subject of basic research, especially in the facilities of the Mote Marine Laboratory in Sarasota. There Dr. Carl Luer has performed groundbreaking studies of skate reproduction, including the first successful artificial insemination in any shark, skate, or ray species.

Range, Distribution, and Habitat Preference

These skates are generally confined to the western North Atlantic. They are abundant around Florida and the eastern Gulf of Mexico but have not been reported in the Bahamas. They may range as far north as Massachusetts, perhaps even to Canada during warmer summer months. They show a preference for shallow water, though trawls have taken animals from more than 1,000 feet (330 m).

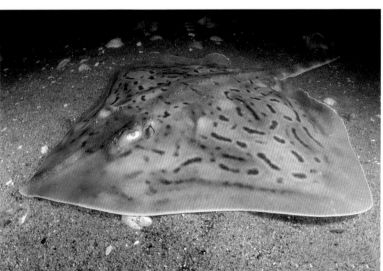

Size, Age, Growth, and Reproduction

Clearnose skates are relatively small animals. Their disks may reach a size of around 19 inches (50 cm) and they may attain an overall length of 33 inches (85 cm). Males and females become sexually mature at a length of almost 30 inches (80 cm).

Skates are among the few elasmobranch species that lay eggs. Only a few sharks reproduce in this manner, but all skates produce egg cases that are shed by the female and develop externally. Mating occurs in the same manner as in other species, with the male passing sperm into the female through his claspers. Sperm may be stored in the female until eggs are produced. Once the eggs have been fertilized, they are covered with a tough egg case until the eggs are shed and embryonic development begins. The young feed on yolk stored in the egg case until they are born. Young clearnose skates are about 5 inches long (13.7 cm) when they hatch. Their disk width is approximately 3.5 to 4.0 inches (9.4 cm).

The skate has become a useful tool for understanding the process of fertilization. It has been used by Dr. Luer for experiments using artificial insemination to examine the feasibility of culture of these marine animals. These techniques may be particularly useful for captive facilities. If the techniques prove successful for other species, the potential for captive breeding could eliminate the need for wild capture, particularly for display animals. In the most optimistic future view, such techniques might hold promise for replenishing depleted stocks of overfished species.

Dr. Luer's studies at Mote Marine Laboratory have been able to follow the entire course of skate embryonic development. The project has produced incredible photographs documenting the various stages of development up to and including egg hatching 12 weeks after the eggs are laid. This is probably the most complete and detailed study of the

The process of embryonic development in the clearnose skate has been studied in detail by Dr. Carl Luer of Mote Marine Laboratory in a delicate process. The skate at four weeks is not yet recognizable, as its development is in its earliest stages. *Photograph courtesy Carl Luer. Used with permission.*

developmental process for any elasmobranch species.

Food and Feeding

Clearnose skates feed upon crustaceans, marine worms, bivalve mollusks such as clams and scallops, squid, and small fish.

Behavior and Interactions with Humans

As bottom-dwellers (benthic organisms), skates have been assumed to move through the water just as most rays do, using their wings to glide along the bottom. But skates are also known to employ a hopping maneuver called punting. It is best described by Laura Macesic and Steve Kajiura, authors of a paper that describes this bizarre swimming adaptation:

> While keeping the rest of the body motionless, the skate's pelvic fins are planted into the substrate and then retracted caudally, which thrusts the body forward. . . . This form of locomotion is not confined to the

At seven weeks, the wings of the clearnose skate have developed, and it looks more like a skate. *Photograph courtesy Carl Luer. Used with permission.*

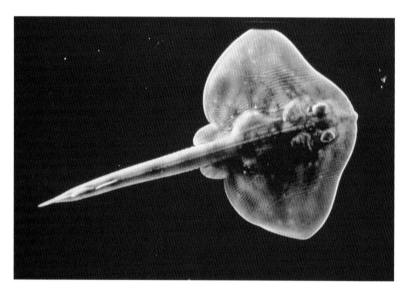

At ten weeks the almost fully formed skate appears as a miniature of the adult. *Photograph courtesy Carl Luer. Used with permission.*

skates, but is found across a range of phylogenetically and morphologically diverse batoid species. However, only the clearnose skate, *Raja eglanteria,* and the lesser electric ray, *Narcine brasiliensis,* performed "true punting," in which only the pelvic fins were engaged. The yellow stingray, *Urobatis jamaicensis,* and the Atlantic stingray, *Dasyatis sabina,* performed "augmented punting," in which pectoral fin movement was also used to generate thrust. Despite this supplemental use of pectoral fins, the augmented punters failed to exceed the punting capabilities of the true punters. The urobatid and the true punters all punted approximately half their disc length per punt, whereas the dasyatid punted a significantly shorter distance. The skate punted significantly faster than the other species. (Macesic and Kajiura 2010, 1219)

Punting may be akin to pouncing and may allow these otherwise slower-moving creatures a mechanism for a surprise attack on unsuspecting prey. The reader is left to ponder the true meaning of a hopping skate or ray.

There are seldom human encounters with clearnose skates. The skates may be seen by waders or beachgoers, and occasionally by divers. The thorns on the skate's dorsal side may be prickly if stepped on, but the absence of a stinging spine means that injuries are minimally severe. Any puncture wound requires cleaning and attention, but wounds from clearnose skates are almost unheard of.

Conservation and Management Status

Clearnose skates are not the target of any directed fishery. Like many other skate and ray species, they are often caught in trawls as bycatch. Analysis of trawling data over time has shown some decrease in population sizes, though no explanations have been given to explain the decreases. The IUCN lists the clearnose skate as a species of "least concern" but notes that the data for this species is sadly out of date.

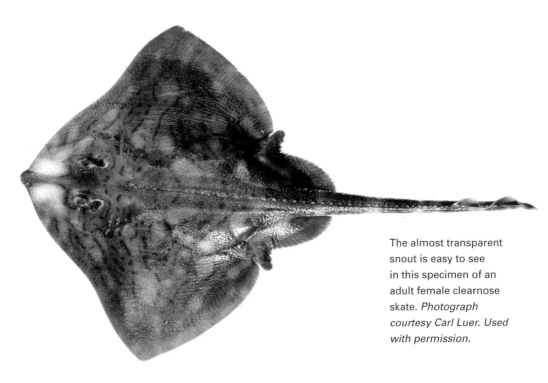

The almost transparent
snout is easy to see
in this specimen of an
adult female clearnose
skate. *Photograph
courtesy Carl Luer. Used
with permission.*

Cownose Ray

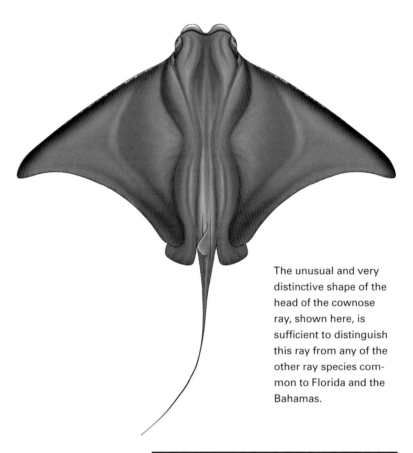

The unusual and very distinctive shape of the head of the cownose ray, shown here, is sufficient to distinguish this ray from any of the other ray species common to Florida and the Bahamas.

Common Names: cownose ray, double-head
Scientific Name: *Rhinoptera bonasus*
Identifying Features: The strangely shaped head and snout make this ray readily identifiable. Its common name is derived from its supposed resemblance to a cow. It is a bulky ray, very square and blocky in its shape and appearance. The wings taper to a sharp triangular shape. Its deep brown to reddish-brown color further distinguishes it from other rays, though its unique head and snout structure make it almost unmistakable.
Frequency: less common

Visitors to any zoo or aquarium where a stingray lagoon or a stingray petting zoo exists have undoubtedly had actual hands-on experiences with cownose rays. They are common animals in these touch-tank encounters. The captive rays are often fed by patrons, who purchase handfuls of small fish and allow the rays to suck them from their open palms. The squeals of youngsters who feel the strange mouth as it nibbles away at the food removes some of the mystery from these animals, even for many visitors who are not particular fans of zoos, aquariums, or other captive facilities. The barbs have been removed from the rays in these pools, to avoid the potential risk of a sting, and the entire experience becomes a friendly interaction with an otherwise unfamiliar animal.

A close relative of eagle rays, cownose rays are often seen in large schools and, when not startled, swim with the same graceful gliding movements as their cousins.

Range, Distribution, and Habitat Preference

Cownose rays are found throughout the western North Atlantic. They are common in northern Florida and the Gulf of Mexico. They may also be seen in southern Florida and the Bahamas, probably more as transients during their migrations than as day-to-day residents. They are considered to be a more pelagic species than the stingrays and eagle rays, but they often venture close to shore. They are also known to reside in some locations for extended periods, even, according to some studies, staying in a small estuary in southwestern Florida for periods as long as 102 days. These studies tracked cownose rays with special transmitters and were able to show that the animals do not always make extensive migrations. In fact, year-round presence in Charlotte Harbor in southwestern Florida may mean that some groups of cownose rays show little or limited migratory

Cownose rays are a familiar sight to anyone who visits large aquariums where a ray-encounter pool is an attraction. Cownose rays are generally the most featured species and are commonly fed by people visiting these exhibits.

behavior. The same studies also revealed that they are tolerant of brackish waters.

Size, Age, Growth, and Reproduction

Cownose rays can reach sizes of around 41 inches (107 cm) disk width. Sexual maturity is reached when the animals are about 33 to 35 inches (85–90 cm), though this size may vary depending on where they live. These figures were determined for animals from the Chesapeake Bay. Another set of data from the Gulf of Mexico found sexually mature cownose rays at disk widths of 25 to 27 inches (65–70 cm). Differences in diet or environment may explain the variability. The age when they attain sexual maturity is between 4 and 5 years. They are thought to live for as long as 13 years.

Very little is known about reproduction in cownose rays. Mating has been seen, though details have not been provided. The size at which sexual maturity is reached varies with location, and the gestation period may also be influenced by where females spend their pregnancy. In some areas they are thought to carry young for a year, while in other areas,

5 to 7 months has been observed. The young are born at a size of 8 to 13 inches (20–38 cm) disk width, and they develop without a placenta. Litters commonly contain only one embryo, but some examinations have shown up to six in a litter.

Food and Feeding

As benthic feeders, cownose rays may find that shallower water provides a greater array of food choices. Bivalve mollusks (such as clams and scallops) are a major component of their diet, but soft-bodied crustaceans are also consumed. Their particular jaw structure and feeding mechanisms suit them well for excavating prey from bottom sands and explain their particular preferences for mollusks and crustaceans.

Behavior and Interactions with Humans

Photographs of gigantic schools of cownose rays during migration are commonplace, and the usual questions apply to the movements of these rays. Where are they going? Why do they travel in large schools? What is guiding

Migrating schools of cownose rays are frequently encountered offshore. These schools may number in the hundreds as the rays move along their migration paths. Recent scientific studies using satellite tracking have shown that cownose rays may also prefer some nearshore areas and may remain residents in these areas for extended periods of time.

their movements? Are they following a food source? Is the schooling a prelude to mating? Are they traveling to a nursery grounds?

Cownose rays seldom interact with humans. They are not regarded as a threat, and the only major consequence of an interaction may be a sting from their barbs. While these rays feed on the bottom, they seldom rest on the bottom. Stepping on one is highly unlikely. Accidental capture in a net or a trawl and subsequent attempts to disentangle the ray from a net may be the most likely way to become injured.

Conservation and Management Status

There does not seem to be a large fishery for cownose rays, though they are taken by artisanal fisheries in some parts of their range. As is usual for many ray species, they are often taken as bycatch by nets or trawls. Since they have little commercial value, and since at least some consumers say their meat tastes terrible, little commercial exploitation has occurred. They are listed by the IUCN as "near threatened," as a caution to monitor nearshore fisheries in parts of their range where data reporting is scarce, to ensure stability of local populations in those areas.

The cownose ray may, in fact, be an unusual species from a commercial standpoint. Cownose rays were thought by some commercial fishers, particularly oyster fishers in the Chesapeake Bay, to pose a significant competitive risk to commercial bivalve mollusk fisheries. People were encouraged to harvest rays recreationally, and a commercial industry was encouraged, to reduce the impact on these commercial mollusk fisheries. Comprehensive studies in recent years, however, have shown such predation by cownose rays not to be the case. The cownose ray's bad reputation among some commercial fishers may be undeserved, and many animals were probably killed as pests when they actually posed no threat at all.

Mantas and Mobulid Rays: The "Devil" Rays

Common Names: manta, manta ray, devil ray, Atlantic devil ray
Scientific Names: *Manta* spp., *Mobula* spp.
Identifying Features: These are large rays with triangular wings; the modified fins on the front of the head form cephalic (head) fins.
Frequency: less common

Manta and mobulid rays, more commonly called devil rays, are among the greatest air-borne acrobats of the seas. Closely related to eagle rays, they also share the love of high-flying antics, reportedly reaching heights of 25 to 30 feet (7–9 m) as they leap, often in groups, for reasons not well understood. Their graceful leaps are slightly tainted by their less-than-graceful landings, when they splash down with awkward belly flops, probably frightening any potential predators that might have prompted their leaps in the first place. They are also among the most graceful swimmers, often engaging in what appear to be choreographed underwater ballets. The tiniest flicks of their wing tips result in barrel rolls and twists and turns that seem impossible for such a large animal.

Manta rays and devil rays are large, filter-feeding rays that are best recognized by the strangely modified fins that are part of the structure of the head. Because of their similarity in appearance and very similar ecologies, they are treated as one large group in this account.

The projections from the front of the head are actually modified pectoral fins called cephalic fins ("cephalo" refers to the head or skull). They are thought to function as scoops

Few rays show more grace as they swim effortlessly through the water than manta rays.

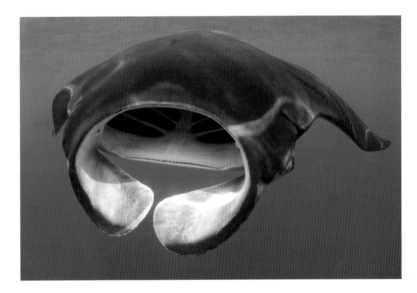

The odd appearance of the manta head and the folded, flaplike appendages that aid in the filter feeding process have inspired names like devil fish or devil ray for the manta and mobulid rays. Such names seem at odds with the benign and gentle nature of a filter-feeding fish.

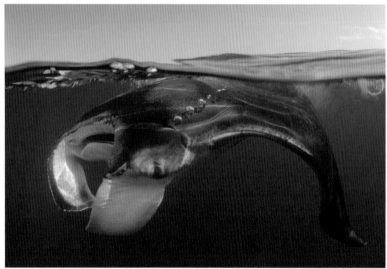

The anterior view of the Caribbean manta (*top*) shows the cephalic fins, often thought of as devil's horns by imaginative storytellers. The side view (*bottom*) shows the flaps distended as the animal cruises the surface waters, where plankton and small fish and crustaceans are abundant.

or funnels that help to channel food into the rays' large mouths when they are feeding on masses of plankton. These fins are generally folded and face forward until the animals feed. Then the fins are unfolded and help to direct the water flow into their mouths. The mouth of a manta ray is on the very front of the head; a mobulid's mouth is more underneath the head.

The classification of these rays is unsettled. They are closely related to eagle rays, but there is continuing debate among taxonomists regarding exactly how many species exist and precisely where they fit in the overall classification of rays. Most current schemes recognize two species of mantas and nine species of devil rays. This account summarizes in very general terms what is known about these species, though not all of them are found in Florida and Bahamian waters.

Range, Distribution, and Habitat Preference

Mantas are worldwide in their distribution, preferring tropical and temperate waters. Devil rays are also common to tropical and temperate waters. Both groups are pelagic and common to open waters near and around continental shelves. However, they are occasionally found near shores and around reef systems. They may be seen both in schools and as solitary individuals.

Size, Age, Growth, and Reproduction

Manta rays are the largest ray species in the world. Some individuals have been measured with a wing span of up to 30 feet (9 m) from wing tip to wing tip, though the average size is somewhat less. Devil rays are considerably smaller, reaching up to 4 feet (120 cm) from wing tip to wing tip.

Because these animals are so large and spend a great deal of their time in open ocean waters, very little is known about growth and aging. Maximum sizes can reach huge proportions for mantas, and they may live for more than 20 years. Some estimates place their life expectancies as high as 30 to 40 years. Nothing is known of the growth rate or life expectancy of the Atlantic devil ray, common in Florida and Bahamas waters.

Mantas and devil rays give birth to one to two live young. Embryos consume yolk until it has been totally absorbed, and then they feed upon secretions produced by the uterine lining. Young mantas may be close to 4 feet (120 cm) at birth. The gestation period is not known. A similar mode of reproduction occurs in devil rays. The litter size is also one to two, and size at birth may be around 20 to

22 inches (55–60 cm). The gestation period for devil rays is also unknown.

Food and Feeding

All of these large rays are filter feeders. Stomach contents of commercially captured and killed individuals typically show large masses of planktonic organisms and other schooling animals, including large masses of small shrimps and small fish.

Interaction with Humans

Interaction with humans is minimal and generally benign. Divers have been fortunate to observe these large rays swimming over reefs, and the rays seem to tolerate underwater photographers. Mantas seem much less threatened than devil rays, who generally scurry away from divers. Many boaters and

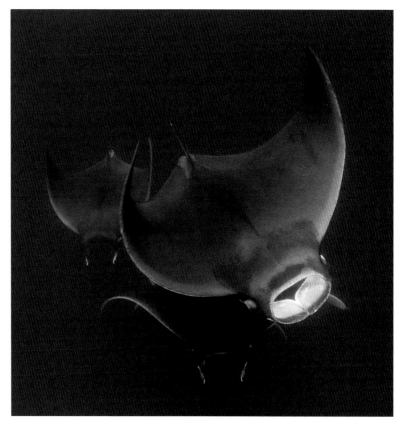

beachgoers have been entertained by leaping rays that may wander nearshore. Other than the sheer size of some individuals, these rays pose no threat to humans.

Conservation and Management Status

Mantas and devil rays are highly fished by commercial enterprises. Their meat is consumed as food, and their livers are processed to extract the oils as medicine. The gill rakers, structures that assist in food removal during the straining of water by these filter feeders, are considered valuable in Asian markets for medicinal purposes.

The rays have been netted, trawled, and caught as bycatch. They are large and slow and are regarded as relatively easy targets with massive weights that can fetch a significant price on weight alone. The markets more heavily target mantas, but devil rays also appear as bycatch. The impact of devil

The folded position of the cephalic fins is easily seen in this anterior view of a mobulid ray. The mouth is less visible in mobulids than in rays because in mobulids it is placed toward the underside of the animal.

A school of mobulas work their way through the water column. In most members of this school, the fins are extended, but they can be seen in the folded position in one animal near the upper right portion of the photograph.

The often huge size of the Atlantic manta can be appreciated in this image, though the distortion from the use of wide-angle lenses somewhat exaggerates the relative size compared to the diver in the background.

The grace and beauty of mantas can be best appreciated by diving with them, though any encounter with these largest of the rays will be memorable.

ray fishing is largely unknown, and devil rays are listed as "data deficient" by the IUCN. Mantas, which are more heavily affected by fishing, are listed as "vulnerable." Many countries have banned fishing efforts for mantas and have instituted heavy fines for violating the laws. Both groups are now protected from international trade by CITES listings.

Roughtail Stingray

Common Name: roughtail stingray
Scientific Name: *Dasyatis centroura*
Identifying Features: The roughtail stingray, as the name suggests, has a tail that is characterized by many rows of thorny spines, making this stingray easy to distinguish from its close cousins in Florida and the Bahamas, the Atlantic stingray and the southern stingray. The tail of the roughtail stingray may be twice as long as the body. These rays can reach very large sizes. They are dark brown in color with a black tail.
Frequency: less common

The roughtail stingray is the largest member of the whiptail stingray group, to which all three of these species belong.

Range, Distribution, and Habitat Preference

Roughtail stingrays are commonly found in the Atlantic and the Mediterranean. Along the US coast, they range from Massachusetts through Florida to the Bahamas and into the Gulf of Mexico. They are found at depths up to 900 feet (274 m). These are among the most extreme depths known for stingrays. Roughtail stingrays are also periodically found nearer the coast, including in bays and estuaries, even penetrating into brackish water. They prefer sandy and muddy bottoms and generally seek warmer waters.

These stingrays are not as commonly encountered in shallow coastal waters in Florida and the Bahamas as the Atlantic and southern stingrays. With their preference for deeper waters and muddy and sandy bottoms, habitats that are not particularly attractive to boaters, swimmers, fishers, or

Roughtail stingrays are seldom encountered in areas with high human presence. They tend to prefer deeper waters and seldom occur inshore, unlike the more common southern stingrays and yellow rays.

Though the roughtail stingray is not as common as other ray species in Florida and Bahamian waters, an accidental encounter with the thorny spines on its dorsal surface and tail could lead to serious puncture wounds.

divers, they are generally not found in areas that humans prefer.

Size, Age, Growth, and Reproduction

As the largest whiptail stingray, roughtails have been measured with disk widths over 8.5 feet (260 cm). An animal of this size would weigh around 640 pounds (290 kg). Such a size is most likely an extreme example. They more commonly average around 7 feet (210–220 cm) disk width.

Roughtail stingrays reach sexual maturity at a disk width of 31 inches (80 cm) for males and between 24 and 40 inches (60–100 cm) for females. Embryos are born live after developing internally, as occurs in most other stingrays. They are nourished by yolk until it has been completely reabsorbed and then rely on fluids produced by the uterus of the mother. There is no placental attachment to the mother.

Litter sizes range from two to six embryos. Gestation may range from 4 months to 10 months, depending on location. Whether they reproduce more than once a year is unknown. The newborn rays are almost 14 inches (34–37 cm) in disk width in some locations. But in warmer water, the gestation

Roughtail stingrays can exceed 7 feet in diameter and weigh hundreds of pounds; they are one of the largest stingrays.

period may be shorter and the size of the newborns correspondingly smaller, averaging 4 to 5 inches (13 cm) in disk width. There is no good explanation for these differences. Most biologists assume that differences in water temperature or in diet or food availability may explain the different developmental and reproductive strategies.

Food and Feeding

As might be predicted, the food of the roughtail stingray consists of bottom-dwelling crustaceans, marine worms, some bony fish, and foods similar to those of other stingrays. Squid, crabs, and shrimps form a major part of their diets. They have been characterized as opportunistic, resting on the bottom waiting for prey to appear, with no particular preference for what they eat. They, in turn, are favored prey items of many sharks, particularly hammerheads, which seem to relish any type of stingray.

Interactions with Humans

These rays are not considered to be aggressive. However, the multiple spines on their tails and the thorny spines on their backs could cause severe puncture wounds if a swimmer or beachgoer should accidentally step on one of these rays. Injuries are far more likely to occur to fishers who accidentally catch roughtail stingrays. Since they can reach massive sizes, care should be taken if they are landed, especially during hook removal and release. Commercial fishers who catch these rays in nets and trawls also exercise extreme care when tending to their catches to avoid the dangerous spines.

Conservation and Management Status

Roughtail stingrays are targeted commercially only in a limited portion of their range. There is no significant fishery along the Atlantic coast of the United States or in the Bahamas. They are occasionally taken as bycatch by commercial trawlers, gillnets, and longlines. The meat is edible, and a number of products are marketed for consumption. Ground meat from the roughtail stingray is used as fishmeal, and there is some value to their liver oils. While caution is recommended and continued monitoring of future fishing efforts is suggested, roughtail stingrays are listed as species of "least concern" by the IUCN.

Sawfish

Common Names: smalltooth sawfish, sawfish, saw shark,

Scientific Name: *Pristis pectinata*

Identifying Features: While there are several species of sawfish, recognizing them as sawfish does not require any particular expertise. Their "saw," an extension of the snout (often referred to as the rostrum), is the characteristic structure for which they are named. Toothlike structures line both sides and pose a formidable threat when the animals decide to feed. Distinguishing between the species of sawfish relies on comparison of tooth size, fin placement, and tooth counts. In spite of their strange and almost sharklike appearance, sawfish are actually more closely related to rays than they are to sharks.

Frequency: rare

Now critically endangered, smalltooth sawfish were once more plentiful in the shallow Florida and Bahamian waters. Worldwide efforts to protect the remaining populations of these animals have been enacted in hopes that declining populations can be restored before sawfish become extinct in the wild. *Photograph copyright Grant Johnson. Used with permission. All rights reserved.*

Most of the research I have conducted on nurse shark growth and movement has been carried out in the very shallow waters of the Florida Keys. Many of the islands of the Keys are isolated and remote. The channels leading to them are shallow and not marked by aids to navigation, and they are not visited by boats as frequently as the easier-to-reach islands. These desolate shallows are wonderful places to encounter sharks, especially small nurse sharks. My preference has always been to fish early in the morning, depending on weather and tides, and watch a new day begin in these biologically rich areas.

On one such occasion, in water less than 6 feet (2 m) deep, the still of the early morning and the mirror-calm water near the shoal where I was fishing was abruptly shattered by sheer chaos. Small finger mullet from a gigantic school suddenly took to the air, many fish actually landing in my boat where I was anchored on the shallow flat. The water churned and became frantic, with fish jumping and splashing everywhere. Out of the midst of this incredible sight emerged a huge saw, slashing back and forth through the water, frightening and injuring any fish unable to escape the scythe that was providing a very bad start to their day.

The pandemonium continued for 5 to 10 minutes, and then an eerie calm settled in. After a short interval, I was able to watch the feeding of a sawfish that I estimated to be 12 to 14 feet (365–400 cm) long as it cruised around my boat, picking up the injured and maimed fish. It casually moved around the flat for another 10 minutes or so, consuming the pieces and parts of mullet not so fortunate as to escape. Watching any large shark or ray feed is exhilarating, but no experience can ever compare to the sight of this efficient

The rostrum, the anterior portion of the head commonly referred to as the saw of the sawfish, has been prized as a trophy. The reduction in numbers of sawfish has led to stringent protections, so that by law, the animals are to be released immediately and unharmed if they are accidentally captured.

thrashing machine harvesting from a school of bait fish. Such sights are increasingly rare because of the reduced numbers of sawfish, though some fortunate commercial fishing guides are occasionally rewarded with a glimpse of these exciting behaviors.

Range, Distribution, and Habitat Preference

Sawfish are now believed to be restricted to the western Atlantic and the Gulf of Mexico. They occur in Florida and the Bahamas, preferring shallow, coastal warm waters near shore or in bays and estuaries. Studies in southwestern Florida following animals that were tagged with satellite tags showed that they spent most of their time in water less than 35 feet (10 m) deep within a temperature range of 72 to 82°F (22 to 28°C). Sawfish have been occasionally found in deeper water and even in freshwater. These same satellite studies show that they move away from their preferred area for some short distances and times but then return to the same area. Bahamian populations have been found to follow this pattern as well. This implies some site fidelity and may be important for management initiatives to protect these known sites as well as the animals themselves.

Size, Age, Growth, and Reproduction

When an animal, such as the sawfish, is not abundant and is potentially dangerous to capture, the study of the animal's basic biology is complicated. Only recently have capture techniques been developed to allow accurate measurements, placement of satellite tags to track the sawfish's movements, and collection of skin samples for DNA analysis. While sawfish's maximum size has been estimated to be around 24 to 25 feet (7.6 m), more common sizes range from 16 to 18 feet (4.9–5.5 m).

Their growth rates and ages have only recently been estimated. Juveniles may grow 25 to 33 inches (65–85 cm) in their first year and 19 to 25 inches (48–68 cm) in their second year. Based on these growth rates, the best estimates place the size of sexual maturity at somewhere between 10 and 14 feet (300–400 cm). These lengths mean that males are probably approaching sexual maturity at 7 to 8 years old and females at 10 to 12 years old.

Sawfish young are nourished through a yolk sac inside the mother and are born live after a gestation period thought to be around 1 year. The rostrum in young sawfish is covered with a protective layer of tissue, which protects the mother during the process of giving birth. It is also somewhat flexible at birth and later becomes more rigid and laden with teeth. Litter sizes are estimated to range from 15 to 20.

Food and Feeding

Sawfish feed in several ways. The saw may be used to stir up the bottom sands as they seek shrimps and crabs from beneath the surface. They may also slash through schools of fish, including mullet and herring, impaling many on their saw or injuring fish that they later return to feed upon. Most studies of their diets rely on observations of feeding. Few stomach-content analyses have been performed.

Interaction with Humans

Although sawfish are not considered to pose a direct threat to humans, the saw is potentially dangerous. Swimming or diving with sawfish without due regard for the injuries that could arise if a sawfish becomes startled might result in severe injuries. Sawfish that are accidentally hooked by fishers must be released immediately and with a great deal of care for both the fisher and the sawfish. Having them alongside a boat for hook removal is hazardous. Their natural tendency is to shake their head, and at that the saw becomes a potential source of injury.

Conservation and Management Status

The sawfish was once highly sought as a trophy, its broad saw taken for wall hangings, ceremonial displays, and ornaments. Fins were also taken and were regarded as a delicacy for shark-fin soup. Live animals are often displayed in captive facilities, where their unique body form remains a curiosity. Diminishing numbers of sawfish may actually have driven the values higher for items that have become a relative rarity.

The smalltooth sawfish is considered to be "critically endangered" by the IUCN; this is the final stage on the route to "extinct in the wild." Its population is thought to have diminished by more than 95% from fishing pressure and from changes to its preferred shallow, nearshore habitats resulting from large-scale coastal urban development.

It is also included by the Convention on International Trade in Endangered Species on a list that bans international trade in endangered species. Such a listing lacks police power, instead relying on member nations to develop laws to combat illicit trade that threatens the existence of a species.

Spotted Eagle Ray

Common Names: spotted eagle ray, eagle ray, white spotted eagle ray

Scientific Name: *Aetobatus narinari*

Identifying Features: Spotted eagle rays are not round in shape but extremely angular with a body shaped more like a diamond than the typical disk shape of a stingray. Their width may be twice as wide as their length, and their tail can be very long if intact. Stinging spines, often numbering two to six, are found near the base of the tail. The white spots and rings covering the upper surface help to identify this species. The spot patterns seem to be different for each animal. Photographic analysis has shown that these patterns can be used much like fingerprints to identify different individuals. This trait may make it possible to create a worldwide database of patterns and unlock secrets of spotted eagle rays' life histories.

Frequency: common

"Soar like an eagle" may apply equally to the bird of prey and to spotted eagle rays. Seldom is a more graceful animal encountered in the coastal waters of Florida and the Bahamas. These large rays with wing spans that may reach 8 to 9 feet (245–275 cm) glide effortlessly through the flats and reef habitats. The beauty of their glides is matched only by the intricacy of the designs that adorn their body and wings. They are truly among the most beautiful of the rays that inhabit our waters.

Range, Distribution, and Habitat Preference

Spotted eagle rays are found worldwide. They prefer tropical and temperate waters and are found nearshore and in open water. They are also common in areas where coral

Wide and pointed wings, a spotted body, and a blunt nose make identification of the spotted eagle ray simple.

Spotted eagle rays flap their pectoral fins, their wings, as a means of propulsion, whereas stingrays rely more on undulations of the pectoral fins for movement.

reefs are abundant. They may be solitary or may cruise in large schools.

Size, Age, Growth, and Reproduction

Sizes typically encountered off southwestern Florida in the Gulf of Mexico show a range

Often seen in groups swimming above sandy bottoms or reefs, spotted eagle rays seem to fly in formation. This observation may be source of their common name. But their name may also have come from their astounding leaps from the water, in which thy appear to fly like eagles.

from 16 to 80 inches (41–203 cm) disk width and weights up to 231 pounds (105 kg). Males become sexually mature at a size of around 47 inches (127 cm) disk width, at an age of 4 to 6 years. Females mature at the same age, but their size at maturity is unknown. Conflicting studies have estimated life expectancy at between 18 and 34 years; there has been no adequate validation of aging in spotted eagle rays.

Their maximum size has been reported to be a width of almost 10 feet (305 cm) and a total length of 8 feet (2.5 m). If the length of the tail is included, the maximum length may reach 16 feet (490 cm) or more. There are reports of weights as high as 507 pounds (230 kg). The IGFA all-tackle world record lists a weight of 273 pounds (124 kg).

Reproduction in rays follows the same pattern as for sharks. A male must establish a grip on the female's pectoral fins and orient himself so that a clasper can be inserted and sperm passed to the female. Mating has not been well documented in large rays in the wild. My one experience with an amorous pair of spotted eagle rays occurred in the shallow waters surrounding the Content

Keys. While snorkeling along a shallow flat to photograph a spotted juvenile nurse shark, I happened upon a pair of eagle rays. The male, much smaller than the female, had established a grip on the back part of the female's pectoral fin and was attempting to properly position himself for mating. Males are generally smaller than females, and this male was having some difficulty in establishing the correct position. As I watched, the female suddenly detected my presence and immediately took flight. The male, who had either bitten the fin in such a way that he could not let go or was frantic that he might not find another willing female, was holding on for dear life. As they took off in the 3 feet of water surrounding us, the last image I had was of the female frantically slapping the surface in her zest to escape, with the small male still attached to her pectoral fin, twisting and turning on the water surface like some bizarre kite tail. Just another adventure in the mysterious love life of spotted eagle rays.

Spotted eagle rays' eggs develop inside the mother, hatch, and then are born alive. Nourishment is produced by the yolk sac and then from uterine milk produced by the lining of the uterus. The gestation period is thought to be 1 year, after which one to four pups are born. Their average disk width at birth is almost 10 inches (26 cm), ranging from 6 to 14 inches (17–36 cm). It is not known how frequently they reproduce.

Food and Feeding

Spotted eagle rays are equipped with a large crushing jaw that suits them well for their chosen prey. Lobsters, crabs, clams, squid, and octopuses are common food items and can be easily crushed by the animals' powerful jaws. Their strange snout may aid them in probing the mud and sand, where they uncover preferred prey items. Like many other ray species, spotted eagle rays fall prey to large sharks, especially large hammerheads.

Behavior and Interactions with Humans

Although spotted eagle rays are graceful when they glide through the water, perhaps with the exception of interrupted mating behaviors, they are capable of huge bursts of speed and are known as well for their aerial displays. They are frequently seen leaping high out of the water with accompanying twists and turns.

Regrettably, those leaps can sometimes have serious consequences. An elderly man was once stabbed in the heart by an airborne eagle ray that landed in his boat; he was impaled by the barb while attempting to return the animal to the water. Several surgeries were required to remove the spine, but the injuries were not fatal. A woman boating in the Florida Keys was not so fortunate. She was struck by a spotted eagle ray as she sat in the front of a boat cruising at high speed. No puncture wounds were detected, and it is thought that the contact was so forceful that both she and the eagle ray were killed. No explanation is given for these fanciful flights, though some scientists believe they are a way to rid the rays of clinging parasites

or to escape a lurking predator. The rays may even undertake these acrobatic leaps to aid the birthing process. Others believe they do it just because they can.

Spotted eagle rays are shy, though they can be tolerant of divers. Underwater photographs and videos reveal their slow, graceful patrols over coral outcroppings, and such photographs have become useful in identifying the unique patterns of spots that aid in identification of individuals.

Conservation and Management Status

In some parts of the world, fishing pressures for spotted eagle rays are severe and their populations are considered at risk. Most landings of them are as bycatch, however. Spotted eagle rays' slow reproductive rates show that a directed fishery might require continuous monitoring to maintain population health and stability. The IUCN considers them to be "near threatened" and worthy of continued monitoring.

The state of Florida totally protects spotted eagle rays in Florida waters and requires that fishers avoid them whenever possible. They are to be released immediately, unharmed, if they happen to be caught.

The pattern of spots on the dorsal surface of the spotted eagle ray is a key to its identification. Scientists have discovered that no two spotted eagle rays have exactly the same pattern of spots. Pattern-recognition software has been used to identify different individuals. Once an individual's identity has been established, a photograph taken of a spotted eagle ray can be a clue to its identification and help biologists determine its movement patterns and secrets of its life history.

Yellow Stingray

Common Names: yellow stingray, yellow spotted stingray

Scientific Name: *Urobatis jamaicensis*

Identifying Features: The size and markings of yellow stingrays are so distinct that they are seldom confused with other rays in Florida and the Bahamas. Their round shape and distinct color patterns make them a great subject for underwater photography. The patterns of color and markings can also serve to disrupt the yellow stingray's shape and may serve to disguise the animal so that it better blends with the bottom. This protective form of coloration may also provide camouflage to aid its survival when it cannot bury itself under the sand. Indeed, small yellow stingrays over a turtle-grass bed are often indistinguishable from the grass bed itself because of the stingray's coloration.

Frequency: common

Yellow stingrays are extremely common inshore rays. They are frequently seen on or near grass beds and in the sandy area surrounding rocks or reefs. The stingray shuffle, the process of wading through shallow water while dragging the feet through the sand, is a helpful way to avoid stepping on one of the rays and receiving a nasty puncture wound from the barb near the end of the tail.

Yellow stingrays are a common sight on coral reefs and on sand and grass flats near the shorelines of Florida and the Bahamas. Like so many other rays found in the tropics, their preference for shallow water often brings them into contact with waders and swimmers, often quite literally. A wader's "stingray shuffle," walking through the shallow water sliding each foot along the bottom, alerts the stingray that something is approaching so that, hopefully, it flees before being stepped on. Those who use beach seines for baitfish or for conducting classes on nearshore environments must use great caution to avoid trampling on the stingrays. Care must be taken when handling captured stingrays because of their struggles that can result in injury from their venomous barb.

The barb is located more toward the end of the tail than in some other stingray species, and it may be more difficult to avoid when the animals have been captured. They will actively thrust the spine in stabbing movements. Wounds are often severe, with the possibility of extreme pain and secondary infection. Barbs may be periodically shed and regrown, and multiple barbs may be present during these times.

Because these animals are generally very small, they have not been fitted with tracking tags or satellite tags to monitor their long-term movements and determine such biological information as growth and aging. Short-term studies, seldom more than 24 hours in duration, have been conducted, but mostly to determine habitat use and activity spaces. Most of the age and growth information comes from studies of captive animals.

While studies comparing the biology of captive animals and wild animals are often similar, there has always been a certain caution in interpreting these results, because

captive environments generally do not per-fectly mimic conditions in the wild. Never-theless, captive studies are regarded as very important; they provide significant baseline data and serve to focus eventual field efforts.

Range, Distribution, and Habitat Preference

Yellow stingrays are found along the south-eastern coast of the United States, throughout the Keys and the Bahamas, and throughout the Gulf of Mexico. They are most frequently found in shallow water near shorelines, where they bury themselves in the sand. The soft bottoms that they prefer often surround seagrass beds, inshore areas that are highly respected and protected as nursery grounds for many commercially and recreationally valuable species of fish and crustaceans. These rich nursery areas also provide plenty of food for the stingrays. Probably because of the abundance of prey items, seagrass beds are where young stingrays are born. Not only is there ample food to nourish the young, but there is sufficient cover for them to hide and have some protection from potential predators.

Size, Age, Growth, and Reproduction

Yellow stingrays seldom exceed 6 to 8 inches in diameter (15–17 cm). Their overall length from the tip of the head to the tip of the tail may be 13 to 14 inches (35 cm). A measured length of 16 inches (39 cm) seems to be the approximate maximum size. Larger sizes have been reported but are thought to be based on estimates rather than actual mea-surements.

Very little is known of yellow stingrays' growth rates and aging in the wild, though they have been successfully bred and raised in captive facilities. Born at about 6 inches (10 cm), they may grow up to 2 inches (5 cm) in their first year, becoming sexually mature by the end of that year. Analysis of vertebral rings—like the study of the growth rings of

trees—shows that they may live as long as 7 to 8 years.

Yellow stingrays may breed up to twice a year; they give birth in shallow waters, gen-erally near seagrass beds. Embryos are nour-ished by a yolk sac and also ingest nutrient-rich secretions called histotroph (uterine milk) from the mother's uterine lining. The gestation period is 5 to 6 months, and the young are born live at a total length of around 5.5 to 6.0 inches (14–16 cm). Litter sizes range from one to seven. Females may actually be-come pregnant again during development of the first litter, so that overlapping develop-ment occurs. This helps to ensure maximum reproductive success.

Food and Feeding

Most rays that are bottom-dwellers feed predominately on small crustaceans such as shrimps and crabs that bury themselves under the sand. Marine worms, small mol-lusks, and small, slow fish may also be parts of yellow stingrays' diet. Their wings are often used to stir up the bottom sands and ex-pose food items that may be hiding just under the surface.

Yellow stingrays themselves may be tar-geted by several species of larger fish and sharks. Goliath groupers (*Epinephelus itajara*)

Juvenile yellow rays are often lighter in color and may more easily blend in with the bottom. Since they are often consumed by large groupers and many species of sharks, whatever protection they can find may give them a survival advan-tage.

Even when the yellow stingray is exposed and not buried, its mottled coloration serves a protective role in allowing it to blend in with the bottom.

have been named as major predators, and gut contents of several shark species have shown remains of yellow stingrays. Often such sharks have yellow stingray barbs stuck in their mouths.

Conservation and Management Status

There is no known targeted commercial fishery for yellow stingrays, though they do appear as bycatch. The yellow stingray's principal commercial value is for the aquarium trade, where it is frequently on display. The main potential risk to populations is thought to be from continued coastal urban development and the habitat loss associated with this development. The IUCN lists the yellow ray as a species of "least concern."

Appendix
Sharks, Skates, and Rays of the World

The following classification is based on Joseph Nelson's *Fishes of the World,* fourth edition, 2006, with some updates from various experts. It is adapted from Helfman and Burgess 2014 and includes many more species than are described in this book to aid the reader in understanding the relationships between the many shark species found elsewhere in the world.

Class Chondrichthyes (cartilaginous fish)
    Subclass Elasmobranchii (sharklike fish)
        Infraclass Euselachii (sharks and rays)
            Division Neoselachii
                Subdivision Selachii (sharks)
                    Superorder Galeomorphi
    Order Heterodontiformes (8 species, marine):
        Heterodontidae (bullhead, horn sharks)
    Order Orectolobiformes (32 species, marine):
        Parascylliidae (collared carpet sharks)
        Brachaeluridae (blind sharks)
        Orectolobidae (wobbegongs)
        Hemiscylliidae (bamboo sharks)
        Stegostomatidae (zebra shark)
        Ginglymostomatidae (nurse sharks)
        Rhincodontidae (whale shark)
    Order Lamniformes (15 species, marine):
        Odontaspididae (sandtiger sharks)
        Mitsukurinidae (goblin shark)
        Pseudocarchariidae (crocodile shark)
        Megachasmidae (megamouth shark)
        Alopiidae (thresher sharks)
        Cetorhinidae (basking shark)
        Lamnidae (mackerel sharks)
    Order Carcharhiniformes (224 species, mostly marine):
        Scyliorhinidae (catsharks)
        Proscylliidae (finback catsharks)
        Pseudotriakidae (false catsharks)

Leptochariidae (barbeled houndshark)

Triakidae (houndsharks)

Hemigaleidae (weasel sharks)

Carcharhinidae (requiem sharks)

Sphyrnidae (hammerhead sharks)

Superorder Squalomorphi

Order Hexanchiformes (5 species, marine):

Chlamydoselachidae (frill sharks)

Hexanchidae (cow sharks)

Order Echinorhiniformes (2 species, marine):

Echinorhinidae (bramble sharks)

Order Squaliformes (97 species, marine):

Squalidae (dogfish sharks)

Centrophoridae (gulper sharks)

Etmopteridae (lantern sharks)

Somniosidae (sleeper sharks)

Oxynotidae (rough sharks)

Dalatiidae (kitefin sharks)

Order Squatiniformes (15 species, marine):

Squatinidae (angel sharks)

Order Pristiophoriformes (5 species, marine):

Pristiophoridae (saw sharks)

Subdivision Batoidea (skates and rays)

Order Torpediniformes (59 species, marine):

Torpedinidae (torpedo electric rays)

Narcinidae (numbfish)

Order Pristiformes (7 species, marine and freshwater):

Pristidae (sawfish)

Order Rajiformes (285 species, marine):

Rhinidae (bowmouth guitarfish)

Rhynchobatidae (wedgefish)

Rhinobatidae (guitarfish)

Rajidae (skates)

Order Myliobatiformes (183 species, marine and freshwater):

Platyrhinidae (thornbacks)

Zanobatidae (striped panrays)

Hexatrygonidae (sixgill stingrays)

Plesiobatidae (deepwater stingrays)

Urolophidae (round stingrays)

Urotrygonidae (American round stingrays)

Dasyatidae (whiptail stingrays)

Potamotrygonidae (river stingrays)

Gymnuridae (butterfly rays)

Rhinopteridae (cownose rays)

Mobulidae (manta rays, mobulas)

References

General

Carrier, J. C., J. A. Musick, and M. R. Heithaus, eds. 2004. *Biology of Sharks and Their Relatives*. Boca Raton, FL: CRC Press, Taylor & Francis Group, LLC.

———, eds. 2010. *Sharks and Their Relatives II*. Boca Raton, FL: CRC Press, Taylor & Francis Group, LLC.

———, eds. 2012. *Biology of Sharks and Their Relatives*. 2nd ed. Boca Raton, FL: CRC Press, Taylor & Francis Group, LLC.

Castro, J. I. 1983. *The Sharks of North American Waters*. College Station: Texas A&M University Press.

———. 2011. *The Sharks of North America*. New York: Oxford University Press.

Ebert, D. A., S. Fowler, and L. J. V. Compagno. 2013. *Sharks of the World*. Plymouth, UK: Wild Nature Press.

Hamlett, W. C., ed. 1999. *Sharks, Skates, and Rays: The Biology of Elasmobranch Fishes*. Baltimore: Johns Hopkins University Press.

———, ed. 2005. *Reproductive Biology and Phylogeny of Chondrichthyes*. Enfield, NH: Science.

Helfman, G., and G. H. Burgess. 2014. *Sharks: The Animal Answer Guide*. Baltimore: Johns Hopkins University Press.

International Union for the Conservation of Nature (IUCN). 2012. *IUCN Red List Categories and Criteria*. Version 3.1. 2nd ed. Cambridge: IUCN.

———. 2016. Introduction. In The IUCN Red List of Threatened Species, iucnredlist.org/about/introduction.

Musick, J. A. 1999. *Life in the Slow Lane: Ecology and Conservation of Long-Lived Marine Animals*. Bethesda: American Fisheries Society.

Neff, C. and R. Hueter. 2013. Science, policy, and the public discourse of shark "attack": A proposal for reclassifying human–shark interactions. *Journal of Environmental Studies and Sciences* 3:65–73.

Tricas, T. C., and S. H. Gruber, eds. 2001. *The Behavior and Sensory Biology of Elasmobranch Fishes*. Dordrecht: Kluwer Academic.

Selected Shark Species
Atlantic Sharpnose Shark

Bethea, D. M., J. K. Carlson, J. A. Buckel, and M. Satterwhite. 2006. Ontogenetic and site-related trends in the diet of the Atlantic sharpnose shark *Rhizoprionodon terraenovae* from the Northeast Gulf of Mexico. *Bulletin of Marine Science* 78 (2): 287–307.

Carlson, J. K., and I. E. Baremore. 2003. Changes in biological parameters of Atlantic sharpnose shark *Rhizoprionodon terraenovae* in the Gulf of Mexico: Evidence for density-dependent growth and maturity? *Marine and Freshwater Research* 54 (3): 227–234.

Carlson, J. K., M. R. Heupel, D. M. Bethea, and L. D. Hollensead. 2008. Coastal habitat use and residency of juvenile Atlantic sharpnose sharks (*Rhizoprionodon terraenovae*). *Estuaries and Coasts* 31 (5): 931–940.

Castro, J. I., and J. P. Wourms. 1993. Reproduction, placentation, and embryonic development of the Atlantic sharpnose shark, *Rhizoprionodon terraenovae*. *Journal of Morphology* 218 (3): 257–280.

Cortes, E. 1995. Demographic analysis of the Atlantic sharpnose shark, *Rhizoprionodon terraenovae*, in the Gulf-of-Mexico. *Fishery Bulletin* 93 (1): 57–66.

Delorenzo, D. M., D. M. Bethea, and J. K. Carlson. 2015. An assessment of the diet and trophic level of Atlantic sharpnose shark Rhizoprionodon terraenovae. *Journal of Fish Biology* 86 (1): 385–391.

Drymon, J. M., S. P. Powers, and R. H. Carmichael. 2012. Trophic plasticity in the Atlantic sharpnose shark (*Rhizoprionodon terraenovae*) from the north central Gulf of Mexico. *Environmental Biology of Fishes* 95 (1): 21–35.

Gelsleichter, J., J. A. Musick, and S. Nichols. 1999. Food habits of the smooth dogfish, *Mustelus canis*, dusky shark, *Carcharhinus obscurus*, Atlantic sharpnose shark, *Rhizoprionodon terraenovae*, and the sand tiger, *Carcharias taurus,* from the Northwest Atlantic Ocean. *Environmental Biology of Fishes* 54 (2): 205–217.

Gurshin, C. W. D., and S. T. Szedlmayer. 2004. Short-term survival and movements of Atlantic sharpnose sharks captured by hook-and-line in the north-east Gulf of Mexico. *Journal of Fish Biology* 65 (4): 973–986.

Hoffmayer, E. R., W. B. Driggers, L. M. Jones, J. M. Hendon, and J. A. Sulikowski. 2013. Variability in the reproductive biology of the Atlantic sharpnose shark in the Gulf of Mexico. *Marine and Coastal Fisheries* 5 (1): 139–151.

Hoffmayer, E. R., G. R. Parsons, and J. Horton. 2006. Seasonal and interannual variation in the energetic condition of adult male Atlantic sharpnose shark *Rhizoprionodon terraenovae* in the northern Gulf of Mexico. *Journal of Fish Biology* 68 (2): 645–653.

Loefer, J. K., and G. R. Sedberry. 2003. Life history of the Atlantic sharpnose shark (*Rhizoprionodon terraenovae*) (Richardson, 1836) off the southeastern United States. *Fishery Bulletin* 101 (1): 75–88.

Marquez-Farias, J. F., and J. L. Castillo-Geniz. 1998. Fishery biology and demography of the Atlantic sharpnose shark, *Rhizoprionodon terraenovae,* in the southern Gulf of Mexico. *Fisheries Research* 39 (2): 183–198.

Parsons, G. R., and E. R. Hoffmayer. 2005. Seasonal changes in the distribution and relative abundance of the Atlantic sharpnose shark *Rhizoprionodon terraenovae* in the north central Gulf of Mexico. *Copeia* (4): 914–920.

Basking Shark

Cotton, P. A., D. W. Sims, S. Fanshawe, and M. Chadwick. 2005. The effects of climate variability on zooplankton and basking shark (*Cetorhinus maximus*) relative abundance off southwest Britain. *Fisheries Oceanography* 14 (2): 151–155.

Gore, M. A., D. Rowat, J. Hall, F. R. Gell, and R. F. Ormond. 2008. Transatlantic migration and deep mid-ocean diving by basking shark. *Biology Letters* 4 (4): 395–398.

Hoelzel, A. R., M. S. Shivji, J. Magnussen, and M. P. Francis. 2006. Low worldwide genetic diversity in the basking shark (*Cetorhinus maximus*). *Biology Letters* 2 (4): 639–642.

Hoogenboom, J. Lisa, Sarah N. P. Wong, Robert A. Ronconi, Heather N. Koopman, Laurie D. Murison, and Andrew J. Westgate. 2015. Environmental predictors and temporal patterns of basking shark (Cetorhinus maximus) occurrence in the lower Bay of Fundy, Canada. *Journal of Experimental Marine Biology and Ecology* 465:24–32.

Kempster, R. M., and S. P. Collin. 2011. Electrosensory pore distribution and feeding in the basking shark *Cetorhinus maximus* (Lamniformes: Cetorhinidae). *Aquatic Biology* 12 (1): 33–36.

Miller, Peter I., Kylie L. Scales, Simon N. Ingram,

Emily J. Southall, and David W. Sims. 2015. Basking sharks and oceanographic fronts: Quantifying associations in the north-east Atlantic. *Functional Ecology* 29 (8): 1099–1109.

Natanson, L. J., S. P. Wintner, F. Johansson, A. Piercy, P. Campbell, A. De Maddalena, S. J. B. Gulak, et al. 2008. Ontogenetic vertebral growth patterns in the basking shark *Cetorhinus maximus. Marine Ecology Progress Series* 361:267–278.

Priede, I. G., and P. I. Miller. 2009. A basking shark (*Cetorhinus maximus*) tracked by satellite together with simultaneous remote sensing II: New analysis reveals orientation to a thermal front. *Fisheries Research* 95 (2–3): 370–372.

Sims, D. W. 2000. Can threshold foraging responses of basking sharks be used to estimate their metabolic rate? *Marine Ecology—Progress Series* 200:289–296.

———. 2008. Sieving a living: A review of the biology, ecology and conservation status of the plankton-feeding basking shark *Cetorhinus maximus.* In *Advances in Marine Biology,* edited by D. W. Sims, 54:171–220. San Diego: Elsevier Academic Press.

Sims, D. W., A. M. Fox, and D. A. Merrett. 1997. Basking shark occurrence off south-west England in relation to zooplankton abundance. *Journal of Fish Biology* 51 (2): 436–440.

Sims, D. W., and D. A. Merrett. 1997. Determination of zooplankton characteristics in the presence of surface feeding basking sharks *Cetorhinus maximus. Marine Ecology—Progress Series* 158:297–302.

Sims, David W., Emily J. Southall, Victoria A. Quayle, and Adrian M. Fox. 2000. Annual social behaviour of basking sharks associated with coastal front areas. *Proceedings of the Royal Society Biological Sciences Series B* 267 (1455): 1897–1904.

Sims, D. W., E. J. Southall, G. A. Tarling, and J. D. Metcalfe. 2005. Habitat-specific normal and reverse diel vertical migration in the plankton-feeding basking shark. *Journal of Animal Ecology* 74 (4): 755–761.

Sims, D. W., C. D. Speedie, and A. M. Fox. 2000. Movements and growth of a female basking shark re-sighted after a three year period. *Journal of the Marine Biological Association of the United Kingdom* 80 (6): 1141–1142.

Skomal, G. B., S. I. Zeeman, J. H. Chisholm, E. L. Summers, H. J. Walsh, K. W. McMahon, and S. R. Thorrold. 2009. Transequatorial migrations by basking sharks in the western Atlantic Ocean. *Current Biology* 19 (12): 1019–1022.

Southall, E. J., D. W. Sims, M. J. Witt, and J. D. Metcalfe. 2006. Seasonal space-use estimates of basking sharks in relation to protection and political-economic zones in the north-east Atlantic. *Biological Conservation* 132 (1): 33–39.

Valeiras, J., A. Lopez, and M. Garcia. 2001. Geographical seasonal occurrence and incidental fishing captures of basking shark *Cetorhinus maximus* (Chondrichyes: Cetorhinidae). *Journal of the*

Marine Biological Association of the United Kingdom 81 (1): 183–184.

Weihs, D. 1999. No hibernation for basking sharks. *Nature (London)* 400 (6746): 717–718.

Wilson, S. G. 2004. Basking sharks (*Cetorhinus maximus*) schooling in the southern Gulf of Maine. *Fisheries Oceanography* 13 (4): 283–286.

Blacknose Shark

Barreto, R. R., R. P. Lessa, F. H. Hazin, and F. M. Santana. 2011. Age and growth of the blacknose shark, *Carcharhinus acronotus* (Poey, 1860) off the northeastern Brazilian coast. *Fisheries Research* 110 (1): 170–176.

Carlson, J. K., E. Cortes, and A. G. Johnson. 1999. Age and growth of the blacknose shark, *Carcharhinus acronotus,* in the eastern Gulf of Mexico. *Copeia* (3): 684–691.

Carlson, J. K., C. L. Palmer, and G. R. Parsons. 1999. Oxygen consumption rate and swimming efficiency of the blacknose shark, *Carcharhinus acronotus. Copeia* (1): 34–39.

Driggers, W. B., J. K. Carlson, B. Cullum, J. M. Dean, D. Oakley, and G. Ulrich. 2004. Age and growth of the blacknose shark, *Carcharhinus acronotus,* in the western North Atlantic Ocean with comments on regional variation in growth rates. *Environmental Biology of Fishes* 71 (2): 171–178.

Giresi, M., M. A. Renshaw, D. S. Portnoy, and J. R. Gold. 2012. Isolation and characterization of microsatellite markers for the blacknose shark, *Carcharhinus acronotus. Conservation Genetics Resources* 4 (1): 141–145.

Sulikowski, J. A., W. B. Driggers, T. S. Ford, R. K. Boonstra, and J. K. Carlson. 2007. Reproductive cycle of the blacknose shark *Carcharhinus acronotus* in the Gulf of Mexico. *Journal of Fish Biology* 70 (2): 428–440.

Blacktip Shark

Baremore, I. E., and M. S. Passerotti. 2013. Reproduction of the blacktip shark in the Gulf of Mexico. *Marine and Coastal Fisheries* 5 (1): 127–138.

Capape, C., A. A. Seck, Y. Diatta, C. Reynaud, F. Hemida, and J. Zaouali. 2004. Reproductive biology of the blacktip shark, *Carcharhinus limbatus* (Chondrichthyes: Carcharhinidae) off west and north African coasts. *Cybium* 28 (4): 275–284.

Chapman, D. D., B. Firchau, and M. S. Shivji. 2008. Parthenogenesis in a large-bodied requiem shark, the blacktip *Carcharhinus limbatus. Journal of Fish Biology* 73 (6): 1473–1477.

Gardiner, Jayne M., Nicholas M. Whitney, and Robert E. Hueter. 2015. Smells like home: The role of olfactory cues in the homing behavior of blacktip sharks, Carcharhinus limbatus. *Integrative and Comparative Biology* 55 (3): 495–506.

Hendon, J. M., D. M. Koester, E. R. Hoffmayer, W. B. Driggers, and A. M. Cicia. 2013. Occurrence of an intersexual blacktip shark in the northern Gulf of Mexico, with notes on the standardization of classifications for this condition in elasmobranchs. *Marine and Coastal Fisheries* 5 (1): 174–180.

Heupel, M. R., and C. A. Simpfendorfer. 2002. Estimation of mortality of juvenile blacktip sharks, *Carcharhinus limbatus,* within a nursery area using telemetry data. *Canadian Journal of Fisheries and Aquatic Sciences* 59 (4): 624–632.

———. 2005. Quantitative analysis of aggregation behavior in juvenile blacktip sharks. *Marine Biology* 147 (5): 1239–1249.

Legare, Bryan, Jeff Kneebone, Bryan DeAngelis, and Gregory Skomal. 2015. The spatiotemporal dynamics of habitat use by blacktip (Carcharhinus limbatus) and lemon (Negaprion brevirostris) sharks in nurseries of St. John, United States Virgin Islands. *Marine Biology* 162 (3): 699–716.

Ritter, Erich K., and Alexander J. Godknecht. 2000. Agonistic displays in the blacktip shark (*Carcharhinus limbatus*). *Copeia* (1): 282–284.

Tavares, R. 2008. Occurrence, diet and growth of juvenile blacktip sharks, *Carcharhinus limbatus,* from Los Roques Archipelago National Park, Venezuela. *Caribbean Journal of Science* 44 (3): 291–302.

Bonnethead Shark

Bethea, D. M., L. Hale, J. K. Carlson, E. Cortes, C. A. Manire, and J. Gelsleichter. 2007. Geographic and ontogenetic variation in the diet and daily ration of the bonnethead shark, *Sphyrna tiburo,* from the eastern Gulf of Mexico. *Marine Biology* 152 (5): 1009–1020.

Carlson, J. K., and G. R. Parsons. 1997. Age and growth of the bonnethead shark, *Sphyrna tiburo,* from northwest Florida, with comments on clinal variation. *Environmental Biology of Fishes* 50 (3): 331–341.

———. 1999. Seasonal differences in routine oxygen consumption rates of the bonnethead shark. *Journal of Fish Biology* 55 (4): 876–879.

Driggers, W. B., B. S. Frazier, D. H. Adams, G. F. Ulrich, C. M. Jones, E. R. Hoffmayer, and M. D. Campbell. 2014. Site fidelity of migratory bonnethead sharks *Sphyrna tiburo* (L. 1758) to specific estuaries in South Carolina, USA. *Journal of Experimental Marine Biology and Ecology* 459:61–69.

Frazier, B. S., W. B. Driggers, D. H. Adams, C. M. Jones, and J. K. Loefer. 2014. Validated age, growth and maturity of the bonnethead *Sphyrna tiburo* in the western North Atlantic Ocean. *Journal of Fish Biology* 85 (3): 688–712.

Gelsleichter, J., C. A. Manire, N. J. Szabo, E. Cortes, J. Carlson, and L. Lombardi-Carlson. 2005. Organochlorine concentrations in bonnethead sharks (*Sphyrna tiburo*) from four Florida estuaries. *Archives of Environmental Contamination and Toxicology* 48 (4): 474–483.

Gelsleichter, J., B. G. Steinetz, C. A. Manire, and

C. Ange. 2003. Serum relaxin concentrations and reproduction in male bonnethead sharks, *Sphyrna tiburo*. *General and Comparative Endocrinology* 132 (1): 27–34.

Harms, C., T. Ross, and A. Segars. 2002. Plasma biochemistry reference values of wild bonnethead sharks, *Sphyrna tiburo*. *Veterinary Clinical Pathology* 31 (3): 111–115.

Heiden, T. C. K., A. N. Haines, C. Manire, J. Lombardi, and T. J. Koob. 2005. Structure and permeability of the egg capsule of the bonnethead shark, *Sphyrna tiburo*. *Journal of Experimental Zoology Part A—Comparative Experimental Biology* 303A (7): 577–589.

Heupel, M. R., C. A. Simpfendorfer, A. B. Collins, and J. P. Tyminski. 2006. Residency and movement patterns of bonnethead sharks, *Sphyrna tiburo*, in a large Florida estuary. *Environmental Biology of Fishes* 76 (1): 47–67.

Jhaveri, Parth, Yannis P. Papastamatiou, and Donovan P. German. 2015. Digestive enzyme activities in the guts of bonnethead sharks (Sphyrna tiburo) provide insight into their digestive strategy and evidence for microbial digestion in their hindguts. *Comparative Biochemistry and Physiology A—Molecular & Integrative Physiology* 189:76–83.

Kroetz, Andrea M., and Sean P. Powers. 2015. Eating between the lines: Functional feeding response of bonnetheads (Sphyrna tiburo). *Environmental Biology of Fishes* 98 (2): 655–661.

Lessa, R. P., and Z. Almeida. 1998. Feeding habits of the bonnethead shark, *Sphyrna tiburo*, from northern Brazil. *Cybium* 22 (4): 383–394.

Manire, C. A., and L. E. L. Rasmussen. 1997. Serum concentrations of steroid hormones in the mature male bonnethead shark, *Sphyrna tiburo*. *General and Comparative Endocrinology* 107 (3): 414–420.

Manire, C. A., L. E. L. Rasmussen, J. Gelsleichter, and D. L. Hess. 2004. Maternal serum and yolk hormone concentrations in the placental viviparous bonnethead shark, *Sphyrna tiburo*. *General and Comparative Endocrinology* 136 (2): 241–247.

Manire, C. A., L. E. L. Rasmussen, D. L. Hess, and R. E. Hueter. 1995. Serum steroid hormones and the reproductive cycle of the female bonnethead shark, *Sphyrna tiburo*. *General and Comparative Endocrinology* 97 (3): 366–376.

Mara, K. R., P. J. Motta, and D. R. Huber. 2010. Bite force and performance in the durophagous bonnethead shark, *Sphyrna tiburo*. *Journal of Experimental Zoology Part A—Ecological Genetics and Physiology* 313A (2): 95–105.

Marquez-Farias, J. F., J. L. Castillo-Geniz, and M. C. R. de la Cruz. 1998. Demography of the bonnethead shark, *Sphyrna tiburo* (Linnaeus, 1758), in the southeastern Gulf of Mexico. *Ciencias Marinas* 24 (1): 13–34.

Nichols, S., J. Gelsleichter, C. A. Manire, and G. M. Cailliet. 2003. Calcitonin-like immunoreactivity in serum and tissues of the bonnethead shark, *Sphyrna tiburo*. *Journal of Experimental Zoology Part A—Comparative Experimental Biology* 298A (2): 150–161.

Parsons, G. R. 1993. Age determination and growth of the bonnethead shark *Sphyrna tiburo*—a comparison of 2 populations. *Marine Biology* 117 (1): 23–31.

———. 1993. Geographic variation in reproduction between two populations of the bonnethead shark, *Sphyrna tiburo*. *Environmental Biology of Fishes* 38 (1–3): 25–35.

Parsons, G. R., and J. K. Carlson. 1998. Physiological and behavioral responses to hypoxia in the bonnethead shark, *Sphyrna tiburo*: Routine swimming and respiratory regulation. *Fish Physiology and Biochemistry* 19 (2): 189–196.

Ubeda, A. J., C. A. Simpfendorfer, and M. R. Heupel. 2009. Movements of bonnetheads, *Sphyrna tiburo*, as a response to salinity change in a Florida estuary. *Environmental Biology of Fishes* 84 (3): 293–303.

Walker, C. J., J. Gelsleichter, D. H. Adams, and C. A. Manire. 2014. Evaluation of the use of metallothionein as a biomarker for detecting physiological responses to mercury exposure in the bonnethead, *Sphyrna tiburo*. *Fish Physiology and Biochemistry* 40 (5): 1361–1371.

Bull Shark

Brunnschweiler, J. M., N. Queiroz, and D. W. Sims. 2010. Oceans apart? Short-term movements and behaviour of adult bull sharks *Carcharhinus leucas* in Atlantic and Pacific Oceans determined from pop-off satellite archival tagging. *Journal of Fish Biology* 77 (6): 1343–1358.

Carlson, J. K., M. M. Ribera, C. L. Conrath, M. R. Heupel, and G. H. Burgess. 2010. Habitat use and movement patterns of bull sharks *Carcharhinus leucas* determined using pop-up satellite archival tags. *Journal of Fish Biology* 77 (3): 661–675.

Cliff, G., and S. F. J. Dudley. 1991. Sharks caught in the protective gill nets off Natal, South-Africa 4: The bull shark *Carcharhinus-leucas* Valenciennes. *South African Journal of Marine Science* 10:253–270.

Curtis, T. H., D. H. Adams, and G. H. Burgess. 2011. Seasonal distribution and habitat associations of bull sharks in the Indian River Lagoon, Florida: A 30-year synthesis. *Transactions of the American Fisheries Society* 140 (5): 1213–1226.

Daly, R., P. W. Froneman, and M. J. Smale. 2013. Comparative feeding ecology of bull sharks (*Carcharhinus leucas*) in the coastal waters of the southwest Indian Ocean inferred from stable isotope analysis. *PLoS One* 8 (10). doi.org/10.1371/journal.pone.0078229.

Drymon, J. M., M. J. Ajemian, and S. P. Powers. 2014. Distribution and dynamic habitat use of young bull sharks *Carcharhinus leucas* in a highly stratified northern Gulf of Mexico estuary. *PLoS One* 9 (5): e97124. doi:10.1371/journal.pone.0097124.

Froeschke, J. T., B. F. Froeschke, and C. M. Stinson. 2013. Long-term trends of bull shark (*Carcharhinus leucas*) in estuarine waters of Texas, USA. *Canadian Journal of Fisheries and Aquatic Sciences* 70 (1): 13–21.

Heupel, M. R., and C. A. Simpfendorfer. 2008. Movement and distribution of young bull sharks *Carcharhinus leucas* in a variable estuarine environment. *Aquatic Biology* 1 (3): 277–289.

———. 2011. Estuarine nursery areas provide a low-mortality environment for young bull sharks *Carcharhinus leucas. Marine Ecology Progress Series* 433:237–244.

Heupel, M. R., B. G. Yeiser, A. B. Collins, L. Ortega, and C. A. Simpfendorfer. 2010. Long-term presence and movement patterns of juvenile bull sharks, *Carcharhinus leucas,* in an estuarine river system. *Marine and Freshwater Research* 61 (1): 1–10.

Natanson, L. J., D. H. Adams, M. V. Winton, and J. R. Maurer. 2014. Age and growth of the bull shark in the Western North Atlantic Ocean. *Transactions of the American Fisheries Society* 143 (3): 732–743.

Simpfendorfer, C. A., G. G. Freitas, T. R. Wiley, and M. R. Heupel. 2005. Distribution and habitat partitioning of immature bull sharks (*Carcharhinus leucas*) in a southwest Florida estuary. *Estuaries* 28 (1): 78–85.

Tillett, J., M. G. Meekan, I. C. Field, D. C. Thorburn, and J. R. Ovenden. 2012. Evidence for reproductive philopatry in the bull shark *Carcharhinus leucas. Journal of Fish Biology* 80 (6): 2140–2158.

Werry, J. M., and E. Clua. 2013. Sex-based spatial segregation of adult bull sharks, *Carcharhinus leucas,* in the New Caledonian great lagoon. *Aquatic Living Resources* 26 (4): 281–288.

Werry, J. M., S. Y. Lee, C. J. Lemckert, and N. M. Otway. 2012. Natural or artificial? Habitat-use by the bull shark, *Carcharhinus leucas. PLoS One* 7 (11): e49796. doi:10.1371/journal.pone.0049796.

Yeiser, B. G., M. R. Heupel, and C. A. Simpfendorfer. 2008. Occurrence, home range and movement patterns of juvenile bull (*Carcharhinus leucas*) and lemon (*Negaprion brevirostris*) sharks within a Florida estuary. *Marine and Freshwater Research* 59 (6): 489–501.

Caribbean Reef Shark

Brooks, E. J., J. W. Mandelman, K. A. Sloman, S. Liss, A. J. Danylchuk, S. J. Cooke, G. B. Skomal, D. P. Philipp, D. W. Sims, and C. D. Suski. 2012. The physiological response of the Caribbean reef shark (*Carcharhinus perezi*) to longline capture. *Comparative Biochemistry and Physiology A—Molecular & Integrative Physiology* 162 (2): 94–100.

Brooks, E. J., D. W. Sims, A. J. Danylchuk, and K. A. Sloman. 2013. Seasonal abundance, philopatry and demographic structure of Caribbean reef shark

(*Carcharhinus perezi*) assemblages in the north-east Exuma Sound, the Bahamas. *Marine Biology* 160 (10): 2535–2546.

Chapman, D. D., E. K. Pikitch, E. A. Babcock, and M. S. Shivji. 2007. Deep-diving and diel changes in vertical habitat use by Caribbean reef sharks *Carcharhinus perezi. Marine Ecology Progress Series* 344:271–275.

Driggers, W. B., E. R. Hoffmayer, E. L. Hickerson, T. L. Martin, and C. T. Gledhill. 2011. Validating the occurrence of Caribbean reef sharks, *Carcharhinus perezi* (Poey), (Chondrichthyes: Carcharhiniformes) in the northern Gulf of Mexico, with a key for sharks of the family Carcharhinidae inhabiting the region. *Zootaxa* 2933:65–68.

Garla, R. C., D. D. Chapman, M. S. Shivji, B. M. Wetherbee, and A. F. Amorim. 2006. Habitat of juvenile Caribbean reef sharks, *Carcharhinus perezi,* at two oceanic insular marine protected areas in the southwestern Atlantic Ocean: Fernando de Noronha Archipelago and Atol das Rocas, Brazil. *Fisheries Research* 81 (2–3): 236–241.

Garla, R. C., D. D. Chapman, B. M. Wetherbee, and M. Shivji. 2006. Movement patterns of young Caribbean reef sharks, *Carcharhinus perezi,* at Fernando de Noronha Archipelago, Brazil: The potential of marine protected areas for conservation of a nursery ground. *Marine Biology* 149 (2): 189–199.

Dusky Shark

Dicken, M. L. 2011. Population size of neonate and juvenile dusky sharks *Carcharhinus obscurus* in the Port of Ngqura, South Africa. *African Journal of Marine Science* 33 (2): 255–261.

Garayzar, C. J. V. 1996. Reproduction of *Carcharhinus obscurus* (Pisces: Carcharhinidae) in the Northeast Pacific. *Revista de Biologia Tropical* 44 (1): 287–289.

Govender, A., and S. L. Birnie. 1997. Mortality estimates for juvenile dusky sharks *Carcharhinus obscurus* in South Africa using mark-recapture data. *South African Journal of Marine Science* 18:11–18.

Hoffmayer, E. R., J. S. Franks, W. B. Driggers, J. A. McKinney, J. M. Hendon, and J. M. Quattro. 2014. Habitat, movements and environmental preferences of dusky sharks, *Carcharhinus obscurus,* in the northern Gulf of Mexico. *Marine Biology* 161 (4): 911–924.

Hussey, N. E., I. D. McCarthy, S. F. J. Dudley, and B. Q. Mann. 2009. Nursery grounds, movement patterns and growth rates of dusky sharks, *Carcharhinus obscurus:* A long-term tag and release study in South African waters. *Marine and Freshwater Research* 60 (6): 571–583.

Marshall, H., G Skomal, P. G. Ross, and D. Bernal. 2015. At-vessel and post-release mortality of the dusky (Carcharhinus obscurus) and sandbar (C. plumbeus) sharks after longline capture. *Fisheries Research* 172:373–384.

Natanson, L. J., J. G. Casey, and N. E. Kohler. 1995. Age

and growth estimates for the dusky shark, *Carcharhinus obscurus,* in the western North Atlantic Ocean. *Fishery Bulletin* 93 (1): 116–126.

Natanson, L. J., B. J. Gervelis, M. V. Winton, L. L. Hamady, S. J. B. Gulak, and J. K. Carlson. 2014. Validated age and growth estimates for *Carcharhinus obscurus* in the northwestern Atlantic Ocean, with pre- and post management growth comparisons. *Environmental Biology of Fishes* 97 (8): 881–896.

Natanson, L. J., and N. E. Kohler. 1996. A preliminary estimate of age and growth of the dusky shark *Carcharhinus obscurus* from the South-West Indian Ocean, with comparisons to the western North Atlantic population. *South African Journal of Marine Science* 17:217–224.

Rogers, P. J., C. Huveneers, S. D. Goldsworthy, J. G. Mitchell, and L. Seuront. 2013. Broad-scale movements and pelagic habitat of the dusky shark *Carcharhinus obscurus* off Southern Australia determined using pop-up satellite archival tags. *Fisheries Oceanography* 22 (2): 102–112.

Romine, J. G., J. A. Musick, and G. H. Burgess. 2009. Demographic analyses of the dusky shark, *Carcharhinus obscurus,* in the Northwest Atlantic incorporating hooking mortality estimates and revised reproductive parameters. *Environmental Biology of Fishes* 84 (3): 277–289.

Simpfendorfer, C. A. 2000. Growth rates of juvenile dusky sharks, *Carcharhinus obscurus* (Lesueur, 1818), from southwestern Australia estimated from tag-recapture data. *Fishery Bulletin* 98 (4): 811–822.

Taylor, S. M., and M. B. Bennett. 2013. Size, sex and seasonal patterns in the assemblage of Carcharhiniformes in a sub-tropical bay. *Journal of Fish Biology* 82 (1): 228–241.

Finetooth Shark

Carlson, J. D., E. Cortes, and D. M. Bethea. 2003. Life history and population dynamics of the finetooth shark (*Carcharhinus isodon*) in the northeastern Gulf of Mexico. *Fishery Bulletin* 101 (2): 281–292.

Driggers, W. B., and E. R. Hoffmayer. 2009. Variability in the reproductive cycle of finetooth sharks, *Carcharhinus isodon,* in the northern Gulf of Mexico. *Copeia* (2): 390–393.

Giresi, M., M. A. Renshaw, D. S. Portnoy, and J. R. Gold. 2012. Development and characterization of microsatellite markers for the finetooth shark, *Carcharhinus isodon. Conservation Genetics Resources* 4 (3): 637–643.

Wiley, T. R., and C. A. Simpfendorfer. 2007. Occurrence of *Carcharhinus isodon* (finetooth shark) in Florida Bay. *Southeastern Naturalist* 6 (1): 183–186.

Great White Shark

Anderson, S. D., T. K. Chapple, S. J. Jorgensen, A. P. Klimley, and B. A. Block. 2011. Long-term individual identification and site fidelity of white sharks, *Carcharodon carcharias,* off California using dorsal fins. *Marine Biology* 158 (6): 1233–1237.

Andrews, A. H., and L. A. Kerr. 2015. Validated age estimates for large white sharks of the northeastern Pacific Ocean: Altered perceptions of vertebral growth shed light on complicated bomb Delta C-14 results. *Environmental Biology of Fishes* 98 (3): 971–978.

Bruce, B. D. 1992. Preliminary observations on the biology of the white shark, *Carcharodon carcharias,* in South Australian Waters. *Australian Journal of Marine and Freshwater Research* 43 (1): 1–11.

Bruce, B., and R. Bradford. 2015. Segregation or aggregation? Sex-specific patterns in the seasonal occurrence of white sharks *Carcharodon carcharias* at the Neptune Islands, South Australia. *Journal of Fish Biology* 87 (6): 1355–1370.

De Vos, A., M. J. O'Riain, M. A. Meyer, P. G. H. Kotze, and A. A. Kock. 2015. Behavior of Cape fur seals (*Arctocephalus pusillus pusillus*) in response to spatial variation in white shark (Carcharodon carcharias) predation risk. *Marine Mammal Science* 31 (3): 1234–1251.

Dureuil, M., A. V. Towner, L. G. Ciolfi, and L. A. Beck. 2015. A computer-aided framework for subsurface identification of white shark pigment patterns. *African Journal of Marine Science* 37 (3): 363–371.

Gilbert, J. M., C. Baduel, Y. Li, A. J. Reichelt-Brushett, P. A. Butcher, S. P. McGrath, V. M. Peddemors, L. Hearn, J. Mueller, and L. Christidis. 2015. Bioaccumulation of PCBs in liver tissue of dusky *Carcharhinus obscurus,* sandbar *C. plumbeus* and white *Carcharodon carcharias* sharks from south-eastern Australian waters. *Marine Pollution Bulletin* 101 (2): 908–913.

Gilbert, J. M., A. J. Reichelt-Brushett, P. A. Butcher, S. P. McGrath, V. M. Peddemors, A. C. Bowling, and L. Christidis. 2015. Metal and metalloid concentrations in the tissues of dusky *Carcharhinus obscurus,* sandbar *C. plumbeus* and white *Carcharodon carcharias* sharks from south-eastern Australian waters, and the implications for human consumption. *Marine Pollution Bulletin* 92 (1–2): 186–194.

Goldman, K. J., and S. D. Anderson. 1999. Space utilization and swimming depth of white sharks, *Carcharodon carcharias,* at the South Farallon Islands, central California. *Environmental Biology of Fishes* 56 (4): 351–364.

Gubili, C., C. E. C. Robinson, G. Cliff, S. P. Wintner, E. de Sabata, S. de Innocentiis, S. Canese, D. W. Sims, A. P. Martin, L. R. Noble, and C. S. Jones. 2015. DNA from historical and trophy samples provides insights into white shark population origins and genetic diversity. *Endangered Species Research* 27 (3): 233–241.

Huveneers, Charlie, Dirk Holman, Rachel Robbins, Andrew Fox, John A. Endler, and Alex H. Taylor. 2015. White sharks exploit the sun during predatory approaches. *American Naturalist* 185 (4): 562–570.

Jewell, O. J. D., R. L. Johnson, E. Gennari, and M. N. Bester. 2013. Fine scale movements and activity areas of white sharks (*Carcharodon carcharias*) in Mossel Bay, South Africa. *Environmental Biology of Fishes* 96 (7): 881–894.

Leurs, G., C. P. O'Connell, S. Andreotti, M. Rutzen, and H. Vonk Noordegraaf. 2015. Risks and advantages of using surface laser photogrammetry on free-ranging marine organisms: A case study on white sharks *Carcharodon carcharias. Journal of Fish Biology* 86 (6): 1713–1728.

Moyer, J. K., M. L. Riccio, and W. E. Bemis. 2015. Development and microstructure of tooth histotypes in the blue shark, *Prionace glauca* (Carcharhiniformes: Carcharhinidae) and the great white shark, *Carcharodon carcharias* (Lamniformes: Lamnidae). *Journal of Morphology* 276 (7): 797–817.

Natanson, L. J., and G. B. Skomal. 2015. Age and growth of the white shark, *Carcharodon carcharias,* in the western North Atlantic Ocean. *Marine and Freshwater Research* 66 (5): 387–398.

O'Leary, S. J., K. A. Feldheim, A. T. Fields, L. J. Natanson, S. Wintner, N. E. Hussey, M. S. Shivji, and D. D. Chapman. 2015. Genetic diversity of white sharks, *Carcharodon carcharias,* in the Northwest Atlantic and Southern Africa. *Journal of Heredity* 106 (3): 258–265.

Onate-Gonzalez, E. C., A. Rocha-Olivares, N. C. Saavedra-Sotelo, and O. Sosa-Nishizaki. 2015. Mitochondrial genetic structure and matrilineal origin of white sharks, *Carcharodon carcharias,* in the Northeastern Pacific: Implications for their conservation. *Journal of Heredity* 106 (4): 347–354.

Skomal, G. B., E. M. Hoyos-Padilla, A. Kukulya, and R. Stokey. 2015. Subsurface observations of white shark *Carcharodon carcharias* predatory behaviour using an autonomous underwater vehicle. *Journal of Fish Biology* 87 (6): 1293–1312.

Tinker, M. T., B. B. Hatfield, M. D. Harris, and J. A. Ames. 2016. Dramatic increase in sea otter mortality from white sharks in California. *Marine Mammal Science* 32 (1): 309–326.

Weng, K. C., J. B. O'Sullivan, C. G. Lowe, C. E. Winkler, H. Dewar, and B. A. Block. 2007. Movements, behavior and habitat preferences of juvenile white sharks *Carcharodon carcharias* in the eastern Pacific. *Marine Ecology Progress Series* 338:211–224.

Wintner, S. P., and G. Cliff. 1999. Age and growth determination of the white shark, *Carcharodon carcharias,* from the east coast of South Africa. *Fishery Bulletin* 97 (1): 153–169.

Hammerhead Sharks

Coelho, R., J. Fernandez-Carvalho, S. Amorim, and M. N. Santos. 2011. Age and growth of the smooth hammerhead shark, *Sphyrna zygaena,* in the Eastern Equatorial Atlantic Ocean, using vertebral sections. *Aquatic Living Resources* 24 (4): 351–357.

Drew, M., W. T. White, Dharmadi, A. V. Harry, and C. Huveneers. 2015. Age, growth and maturity of the pelagic thresher Alopias pelagicus and the scalloped hammerhead Sphyrna lewini. *Journal of Fish Biology* 86 (1): 333–354.

Gulak, S. J. B., A. J. de Ron Santiago, and J. K. Carlson. 2015. Hooking mortality of scalloped hammerhead Sphyrna lewini and great hammerhead Sphyrna mokarran sharks caught on bottom longlines. *African Journal of Marine Science* 37 (2): 267–273.

Harry, A. V., W. G. Macbeth, A. N. Gutteridge, and C. A. Simpfendorfer. 2011. The life histories of endangered hammerhead sharks (Carcharhiniformes, Sphyrnidae) from the east coast of Australia. *Journal of Fish Biology* 78 (7): 2026–2051.

Kotas, J. E., V. Mastrochirico, and M. Petrere. 2011. Age and growth of the scalloped hammerhead shark, *Sphyrna lewini* (Griffith and Smith, 1834), from the southern Brazilian coast. *Brazilian Journal of Biology* 71 (3): 755–761.

Loor-Andrade, P., F. Galvan-Magana, F. R. Elorriaga-Verplancken, C. Polo-Silva, and A. Delgado-Huertas. 2015. Population and individual foraging patterns of two hammerhead sharks using carbon and nitrogen stable isotopes. *Rapid Communications in Mass Spectrometry* 29 (9): 821–829.

Lowe, C. G. 2002. Bioenergetics of free-ranging juvenile scalloped hammerhead sharks (*Sphyrna lewini*) in Kane'ohe Bay, O'ahu, HI. *Journal of Experimental Marine Biology and Ecology* 278 (2): 141–156.

Mara, Kyle R., Philip J. Motta, Andrew P. Martin, and Robert E. Hueter. 2015. Constructional morphology within the head of hammerhead sharks (Sphyrnidae). *Journal of Morphology* 276 (5): 526–539.

McComb, D. M., T. C. Tricas, and S. M. Kajiura. 2009. Enhanced visual fields in hammerhead sharks. *Journal of Experimental Biology* 212 (24): 4010–4018.

Nava, P. N., and J. F. Marquez-Farias. 2014. Size at maturity of the smooth hammerhead shark, *Sphyma zygaena,* captured in the Gulf of California. *Hidrobiologica* 24 (2): 129–135.

Passerotti, M. S., J. K. Carlson, A. N. Piercy, and S. E. Campana. 2010. Age validation of great hammerhead shark (*Sphyrna mokarran*), determined by bomb radiocarbon analysis. *Fishery Bulletin* 108 (3): 346–351.

Piercy, A. N., J. K. Carlson, and M. S. Passerotti. 2010. Age and growth of the great hammerhead shark, *Sphyrna mokarran,* in the north-western Atlantic Ocean and Gulf of Mexico. *Marine and Freshwater Research* 61 (9): 992–998.

Piercy, A. N., J. K. Carlson, J. A. Sulikowski, and G. H. Burgess. 2007. Age and growth of the scalloped hammerhead shark, *Sphyrna lewini,* in the northwest Atlantic Ocean and Gulf of Mexico. *Marine and Freshwater Research* 58 (1): 34–40.

Shiffman, David. 2016. It's illegal for anglers to land hammerheads in Florida. Southern Fried Science. www.southernfriedscience.com/its-illegal-for -anglers-to-land-hammerheads-in-florida-its-time -that-media-coverage-pointed-that-out/.

White, W. T., C. Bartron, and I. C. Potter. 2008. Catch composition and reproductive biology of *Sphyrna lewini* (Griffith & Smith) (Carcharhiniformes, Sphyrnidae) in Indonesian waters. *Journal of Fish Biology* 72 (7): 1675–1689.

Lemon Shark

Barker, M. J., S. H. Gruber, S. P. Newman, and V. Schluessel. 2005. Spatial and ontogenetic variation in growth of nursery-bound juvenile lemon sharks, *Negaprion brevirostris:* A comparison of two age-assigning techniques. *Environmental Biology of Fishes* 72 (3): 343–355.

Brooks, E. J., K. A. Sloman, S. Liss, L. Hassan-Hassanein, A. J. Danylchuk, S. J. Cooke, J. W. Mandelman, G. B. Skomal, D. W. Sims, and C. D. Suski. 2011. The stress physiology of extended duration tonic immobility in the juvenile lemon shark, *Negaprion brevirostris* (Poey 1868). *Journal of Experimental Marine Biology and Ecology* 409 (1–2): 351–360.

Brunnschweiler, J. M. 2009. Tracking free-ranging sharks with hand-fed intra-gastric acoustic transmitters. *Marine and Freshwater Behaviour and Physiology* 42 (3): 201–209.

Bullock, R. W., T. L. Guttridge, I. G. Cowx, M. Elliott, and S. H. Gruber. 2015. The behaviour and recovery of juvenile lemon sharks Negaprion brevirostris in response to external accelerometer tag attachment. *Journal of Fish Biology* 87 (6): 1342–1354.

Carlson, J. K., L. F. Hale, A. Morgan, and G. Burgess. 2012. Relative abundance and size of coastal sharks derived from commercial shark longline catch and effort data. *Journal of Fish Biology* 80 (5): 1749–1764.

Chapman, D. D., E. A. Babcock, S. H. Gruber, J. D. Dibattista, B. R. Franks, S. A. Kessel, T. Guttridge, E. K. Pikitch, and K. A. Feldheim. 2009. Long-term natal site-fidelity by immature lemon sharks (*Negaprion brevirostris*) at a subtropical island. *Molecular Ecology* 18 (16): 3500–3507.

Cortes, E., and S. H. Gruber. 1992. Gastric evacuation in the young lemon shark, *Negaprion brevirostris,* under field conditions. *Environmental Biology of Fishes* 35 (2): 205–212.

DeAngelis, B. M., C. T. McCandless, N. E. Kohler, C. W. Recksiek, and G. B. Skomal. 2008. First characterization of shark nursery habitat in the United States Virgin Islands: Evidence of habitat partitioning by two shark species. *Marine Ecology Progress Series* 358:257–271.

De Freitas, R. H. A., R. S. Rosa, B. M. Wetherbee, and S. H. Gruber. 2009. Population size and survivorship for juvenile lemon sharks (*Negaprion brevirostris*) on their nursery grounds at a marine protected area in Brazil. *Neotropical Ichthyology* 7 (2): 205–212.

DiGirolamo, A. L., S. H. Gruber, C. Pomory, and W. A. Bennett. 2012. Diel temperature patterns of juvenile lemon sharks *Negaprion brevirostris,* in a shallow-water nursery. *Journal of Fish Biology* 80 (5): 1436–1448.

Edren, S. M. C., and S. H. Gruber. 2005. Homing ability of young lemon sharks, *Negaprion brevirostris. Environmental Biology of Fishes* 72 (3): 267–281.

Feldheim, K. A., S. H. Gruber, and M. V. Ashley. 2001. Multiple paternity of a lemon shark litter (Chondrichthyes: Carcharhinidae). *Copeia* 2001 (3): 781–786.

———. 2001. Population genetic structure of the lemon shark (*Negaprion brevirostris*) in the western Atlantic: DNA microsatellite variation. *Molecular Ecology* 10 (2): 295–303.

———. 2002. The breeding biology of lemon sharks at a tropical nursery lagoon. *Proceedings of the Royal Society Biological Sciences Series B* 269 (1501): 1655–1661.

Feldheim, K. A., S. H. Gruber, J. D. Dibattista, E. A. Babcock, S. T. Kessel, A. P. Hendry, E. K. Pikitch, M. V. Ashley, and D. D. Chapman. 2014. Two decades of genetic profiling yields first evidence of natal philopatry and long-term fidelity to parturition sites in sharks. *Molecular Ecology* 23 (1): 110–117.

Freitas, R. H. A., R. S. Rosa, S. H. Gruber, and B. M. Wetherbee. 2006. Early growth and juvenile population structure of lemon sharks *Negaprion brevirostris* in the Atol das Rocas Biological Reserve, off north-east Brazil. *Journal of Fish Biology* 68 (5): 1319–1332.

Gruber, Samuel H., Jean R. C. de Marignac, and John M. Hoenig. 2001. Survival of juvenile lemon sharks at Bimini, Bahamas, estimated by mark-depletion experiments. *Transactions of the American Fisheries Society* 130 (3): 376–384.

Guttridge, T. L., S. van Dijk, E. J. Stamhuis, J. Krause, S. H. Gruber, and C. Brown. 2013. Social learning in juvenile lemon sharks, *Negaprion brevirostris. Animal Cognition* 16 (1): 55–64.

Hueter, R. E. 1990. Adaptations for spatial vision in sharks. *Journal of Experimental Zoology* S5:130–141.

Jennings, D. E., S. H. Gruber, B. R. Franks, S. T. Kessel, and A. L. Robertson. 2008. Effects of large-scale anthropogenic development on juvenile lemon shark (*Negaprion brevirostris*) populations of Bimini, Bahamas. *Environmental Biology of Fishes* 83 (4): 369–377.

Kessel, S. T., D. D. Chapman, B. R. Franks, T. Gedamke, S. H. Gruber, J. M. Newman, E. R. White, and R. G. Perkins. 2014. Predictable temperature-regulated residency, movement and migration in a large, highly mobile marine predator (*Negaprion brevirostris*). *Marine Ecology Progress Series* 514:175–190.

Legare, B., J. Kneebone, B. M. DeAngelis, and G.

Skomal. 2015. The spatiotemporal dynamics of habitat use by blacktip (Carcharhinus limbatus) and lemon (Negaprion brevirostris) sharks in nurseries of St. John, United States Virgin Islands. *Marine Biology* 162 (3): 699–716.

Morrissey, J. F., and S. H. Gruber. 1993. Habitat selection by juvenile lemon sharks, *Negaprion brevirostris*. *Environmental Biology of Fishes* 38 (4): 311–319.

Motta, P. J., T. C. Tricas, and R. E. Hueter. 1997. Feeding mechanism and functional morphology of the jaws of the lemon shark *Negaprion brevirostris* (Chondrichthyes, Carcharhinidae). *Journal of Experimental Biology* 200:2765–2780.

Motta, P. J., and C. A. D. Wilga. 1995. Anatomy of the feeding apparatus of the lemon shark, *Negaprion brevirostris*. *Journal of Morphology* 226 (3): 309–329.

Murchie, K. J., E. Schwager, S. J. Cooke, A. J. Danylchuk, S. E. Danylchuk, T. L. Goldberg, C. D. Suski, and D. P. Philipp. 2010. Spatial ecology of juvenile lemon sharks (*Negaprion brevirostris*) in tidal creeks and coastal waters of Eleuthera, the Bahamas. *Environmental Biology of Fishes* 89 (1): 95–104.

Newman, S. P., R. D. Handy, and S. H. Gruber. 2010. Diet and prey preference of juvenile lemon sharks *Negaprion brevirostris*. *Marine Ecology Progress Series* 398:221–234.

O'Connell, C. P., T. L. Guttridge, S. H. Gruber, J. Brooks, J. S. Finger, and P. He. 2014. Behavioral modification of visually deprived lemon sharks (*Negaprion brevirostris*) towards magnetic fields. *Journal of Experimental Marine Biology and Ecology* 453:131–137.

Reeve, A., R. D. Handy, and S. H. Gruber. 2009. Prey selection and functional response of juvenile lemon sharks *Negaprion brevirostris*. *Journal of Fish Biology* 75 (1): 276–281.

Sundstrom, L. F., S. H. Gruber, S. M. Clermont, J. P. S. Correia, J. R. C. de Marignac, J. F. Morrissey, C. R. Lowrance, L. Thomassen, and M. T. Oliveira. 2001. Review of elasmobranch behavioral studies using ultrasonic telemetry with special reference to the lemon shark, *Negaprion brevirostris,* around Bimini Islands, Bahamas. *Environmental Biology of Fishes* 60 (1–3): 225–250.

Wetherbee, B. M., S. H. Gruber, and R. S. Rosa. 2007. Movement patterns of juvenile lemon sharks *Negaprion brevirostris* within Atol das Rocas, Brazil: A nursery characterized by tidal extremes. *Marine Ecology Progress Series* 343:283–293.

White, E. R., J. D. Nagy, and S. H. Gruber. 2014. Modeling the population dynamics of lemon sharks. *Biology Direct* 9:23.

Mako Shark

Abascal, F. J., M. Quintans, A. Ramos-Cartelle, and J. Mejuto. 2011. Movements and environmental preferences of the shortfin mako, *Isurus oxyrinchus,* in the southeastern Pacific Ocean. *Marine Biology* 158 (5): 1175–1184.

Ardizzone, D., G. M. Cailliet, L. J. Natanson, A. H. Andrews, L. A. Kerr, and T. A. Brown. 2006. Application of bomb radiocarbon chronologies to shortfin mako (*Isurus oxyrinchus*) age validation. *Environmental Biology of Fishes* 77 (3–4): 355–366.

Bustamante, C., and M. B. Bennett. 2013. Insights into the reproductive biology and fisheries of two commercially exploited species, shortfin mako (*Isurus oxyrinchus*) and blue shark (*Prionace glauca*), in the south-east Pacific Ocean. *Fisheries Research* 143:174–183.

Campana, S., L. J. Natanson, and S. Myklevoll. 2002. Bomb dating and age determination of large pelagic sharks. *Canadian Journal of Fisheries and Aquatic Sciences* 59 (3): 450–455.

Casey, J. G., and N. E. Kohler. 1992. Tagging studies on the shortfin mako shark (*Isurus oxyrinchus*) in the western North Atlantic. *Australian Journal of Marine and Freshwater Research* 43 (1): 45–60.

Cerna, F., and R. Licandeo. 2009. Age and growth of the shortfin mako (*Isurus oxyrinchus*) in the southeastern Pacific off Chile. *Marine and Freshwater Research* 60 (5): 394–403.

Chang, J. H., and K. M. Liu. 2009. Stock assessment of the shortfin mako shark (*Isurus oxyrinchus*) in the northwest Pacific Ocean using per recruit and virtual population analyses. *Fisheries Research* 98 (1–3): 92–101.

Corrigan, S., D. Kacev, and J. Werry. 2015. A case of genetic polyandry in the shortfin mako Isurus oxyrinchus. *Journal of Fish Biology* 87 (3): 794–798.

Donley, J. A., R. E. Shadwick, C. A. Sepulveda, P. Konstantinidis, and S. Gemballa. 2005. Patterns of red muscle strain/activation and body kinematics during steady swimming in a lamnid shark, the shortfin mako (*Isurus oxyrinchus*). *Journal of Experimental Biology* 208 (12): 2377–2387.

Dono, F., S. Montealegre-Quijano, A. Domingo, and P. G. Kinas. 2015. Bayesian age and growth analysis of the shortfin mako shark Isurus oxyrinchus in the western South Atlantic Ocean using a flexible model. *Environmental Biology of Fishes* 98 (2): 517–533.

Harford, W. F. 2013. Trophic modeling of shortfin mako (*Isurus oxyrinchus*) and bluefish (*Pomatomus saltatrix*) interactions in the western North Atlantic Ocean. *Bulletin of Marine Science* 89 (1): 161–188.

Heist, E. J., J. A. Musick, and J. E. Graves. 1996. Genetic population structure of the shortfin mako (*Isurus oxyrinchus*) inferred from restriction fragment length polymorphism analysis of mitochondrial DMA. *Canadian Journal of Fisheries and Aquatic Sciences* 53 (3): 583–588.

Hurley, P. C. F. 1998. A review of the fishery for pelagic sharks in Atlantic Canada. *Fisheries Research* 39 (2): 107–113.

Joung, S. J., and H. H. Hsu. 2005. Reproduction and embryonic development of the shortfin mako, *Isurus oxyrinchus* Rafinesque, 1810, in the northwestern Pacific. *Zoological Studies* 44 (4): 487–496.

Lopez, S., R. Melendez, and P. Barria. 2009. Feeding of the shortfin mako shark *Isurus oxyrinchus* Rafinesque, 1810 (Lamniformes: Lamnidae) in the Southeastern Pacific. *Revista de Biologia Marina y Oceanografia* 44 (2): 439–451.

MacNeill, M. A., G. B. Skomal, and A. T. Fisk. 2005. Stable isotopes from multiple tissues reveal diet switching in sharks. *Marine Ecology—Progress Series* 302:199–206.

Maia, A., N. Queiroz, H. N. Cabral, A. M. Santos, and J. P. Correia. 2007. Reproductive biology and population dynamics of the shortfin mako, *Isurus oxyrinchus* Rafinesque, 1810, off the southwest Portuguese coast, eastern North Atlantic. *Journal of Applied Ichthyology* 23 (3): 246–251.

Maia, A., N. Queiroz, J. P. Correia, and H. Cabral. 2006. Food habits of the shortfin mako, *Isurus oxyrinchus,* off the southwest coast of Portugal. *Environmental Biology of Fishes* 77 (2): 157–167.

Mollet, H. F., G. Cliff, H. L. Pratt, and J. D. Stevens. 2000. Reproductive biology of the female shortfin mako, *Isurus oxyrinchus* Rafinesque, 1810, with comments on the embryonic development of lamnoids. *Fishery Bulletin* 98 (2): 299–318.

Mucientes, G. R., N. Queiroz, L. L. Sousa, P. Tarroso, and D. W. Sims. 2009. Sexual segregation of pelagic sharks and the potential threat from fisheries. *Biology Letters* 5 (2): 156–159.

Natanson, L. J., N. E. Kohler, D. Ardizzone, G. M. Cailliet, S. P. Wintner, and H. F. Mollet. 2006. Validated age and growth estimates for the shortfin mako, *Isurus oxyrinchus,* in the North Atlantic Ocean. *Environmental Biology of Fishes* 77 (3–4): 367–383.

Preti, A., C. U. Soykan, H. Dewar, R. J. D. Wells, N. Spear, and S. Kohin. 2012. Comparative feeding ecology of shortfin mako, blue and thresher sharks in the California Current. *Environmental Biology of Fishes* 95 (1): 127–146.

Ribot-Carballal, M. C., F. Galvan-Magana, and C. Quinonez-Velazquez. 2005. Age and growth of the shortfin mako shark, *Isurus oxyrinchus,* from the western coast of Baja California Sur, Mexico. *Fisheries Research* 76 (1): 14–21.

Semba, Y., I. Aoki, and K. Yokawa. 2011. Size at maturity and reproductive traits of shortfin mako, *Isurus oxyrinchus,* in the western and central North Pacific. *Marine and Freshwater Research* 62 (1): 20–29.

Semba, Y., H. Nakano, and I. Aoki. 2009. Age and growth analysis of the shortfin mako, *Isurus oxyrinchus,* in the western and central North Pacific Ocean. *Environmental Biology of Fishes* 84 (4): 377–391.

Sepulveda, C. A., S. Kohin, C. Chan, R. Vetter, and J. B. Graham. 2004. Movement patterns, depth preferences, and stomach temperatures of free-swimming juvenile mako sharks, *Isurus oxyrinchus,* in the Southern California Bight. *Marine Biology* 145 (1): 191–199.

Wells, R. J. D., S. E. Smith, S. Kohin, E. Freund, N. Spear, and D. A. Ramon. 2013. Age validation of juvenile shortfin mako (*Isurus oxyrinchus*) tagged and marked with oxytetracycline off southern California. *Fishery Bulletin* 111 (2): 147–160.

Wood, A. D., J. S. Collie, and N. E. Kohler. 2007. Estimating survival of the shortfin mako *Isurus oxyrinchus* (Rafinesque) in the north-west Atlantic from tag-recapture data. *Journal of Fish Biology* 71 (6): 1679–1695.

Wood, A. D., B. M. Wetherbeez, F. Juanes, N. E. Kohler, and C. Wilga. 2009. Recalculated diet and daily ration of the shortfin mako (*Isurus oxyrinchus*), with a focus on quantifying predation on bluefish (*Pomatomus saltatrix*) in the northwest Atlantic Ocean. *Fishery Bulletin* 107 (1): 76–88.

Nurse Shark

Carrier, J. C., and C. A. Luer. 1990. Growth rates in the nurse shark, *Ginglymostoma cirratum. Copeia* 1990 (3): 686–692.

Carrier, J. C., F. L. Murru, M. T. Walsh, and H. L. Pratt. 2003. Assessing reproductive potential and gestation in nurse sharks (*Ginglymostoma cirratum*) using ultrasonography and endoscopy: An example of bridging the gap between field research and captive studies. *Zoo Biology* 22 (2): 179–187.

Carrier, Jeffrey C., Harold L. Pratt Jr., and Linda K. Martin. 1994. Group reproductive behaviors in free-living nurse sharks, *Ginglymostoma cirratum. Copeia* 1994 (3): 646–656.

Casper, B. M., and D. A. Mann. 2006. Evoked potential audiograms of the nurse shark (Ginglymostoma cirratum) and the yellow stingray (Urobatis jamaicensis). *Environmental Biology of Fishes* 76 (1): 101–108.

Castro, Jose I. 2000. The biology of the nurse shark, *Ginglymostoma cirratum,* off the Florida east coast and the Bahamas Islands. *Environmental Biology of Fishes* 58 (1): 1–22.

Hannan, K. M., W. B. Driggers, D. S. Hanisko, L. M. Jones, and A. B. Canning. 2012. Distribution of the nurse shark, *Ginglymostoma cirratum,* in the northern Gulf of Mexico. *Bulletin of Marine Science* 88 (1): 73–80.

Matott, M. P., P. J. Motta, and R. E. Hueter. 2005. Modulation in feeding kinematics and motor pattern of the nurse shark *Ginglymostoma cirratum. Environmental Biology of Fishes* 74 (2): 163–174.

Motta, P. J., and C. D. Wilga. 1999. Anatomy of the feeding apparatus of the nurse shark, *Ginglymostoma cirratum. Journal of Morphology* 241 (1): 33–60.

Rouse, N. 1985. Nurse sharks' mating ballet. *Sea Frontiers* 31:54–57.

Oceanic Whitetip Shark

Brodziak, J., and W. A. Walsh. 2013. Model selection and multimodel inference for standardizing catch rates of bycatch species: A case study of oceanic whitetip shark in the Hawaii-based longline fishery. *Canadian Journal of Fisheries and Aquatic Sciences* 70 (12): 1723–1740.

Howey-Jordan, L. A., E. J. Brooks, D. L. Abercrombie, L. K. B. Jordan, A. Brooks, S. Williams, E. Gospodarczyk, and D. D. Chapman. 2013. Complex movements, philopatry and expanded depth range of a severely threatened pelagic shark, the oceanic whitetip (*Carcharhinus longimanus*) in the western North Atlantic. *PLoS One* 8 (2): e56588. doi:10.1371/journal.pone.0056588.

Lessa, R., R. Paglerani, and F. M. Santana. 1999. Biology and morphometry of the oceanic whitetip shark, *Carcharhinus longimanus* (Carcharhinidae), off north-eastern Brazil. *Cybium* 23 (4): 353–368.

Madigan, D. J., E. J. Brooks, M. E. Bond, J. Gelsleichter, L. A. Howey, D. L. Abercrombie, A. Brooks, and D. D. Chapman. 2015. Diet shift and site-fidelity of oceanic whitetip sharks *Carcharhinus longimanus* along the Great Bahama Bank. *Marine Ecology Progress Series* 529:185–197.

Tambourgi, M. R. D., F. H. V. Hazin, P. G. V. Oliveira, R. Coelho, G. Burgess, and P. C. G. Roque. 2013. Reproductive aspects of the oceanic whitetip shark, *Carcharhinus longimanus* (Elasmobranchii: Carcharhinidae), in the equatorial and southwestern Atlantic Ocean. *Brazilian Journal of Oceanography* 61 (2): 161–168.

Tolotti, M. T., P. Bach, F. Hazin, P. Travassos, and L. Dagorn. 2015. Vulnerability of the oceanic whitetip shark to pelagic longline fisheries. *PLoS One* 10 (10): e0141396. doi:10.1371/journal.pone.0141396.

Tolotti, M. T., P. Travassos, F. L. Fredou, C. Wor, H. A. Andrade, and F. Hazin. 2013. Size, distribution and catch rates of the oceanic whitetip shark caught by the Brazilian tuna longline fleet. *Fisheries Research* 143:136–142.

Sandbar Shark

Andrews, A. H., L. J. Natanson, L. A. Kerr, G. H. Burgess, and G. M. Cailliet. 2011. Bomb radiocarbon and tag-recapture dating of sandbar shark (*Carcharhinus plumbeus*). *Fishery Bulletin* 109 (4): 454–465.

Baremore, I. E., and L. F. Hale. 2012. Reproduction of the sandbar shark in the western North Atlantic Ocean and Gulf of Mexico. *Marine and Coastal Fisheries* 4 (1): 560–572.

Brewster-Geisz, K. K., and T. J. Miller. 2000. Management of the sandbar shark, *Carcharhinus plumbeus:* Implications of a stage-based model. *Fishery Bulletin* 98 (2): 236–249.

Casey, J. G., and L. J. Natanson. 1992. Revised estimates of age and growth of the sandbar shark (*Carcharhinus plumbeus*) from the western North-Atlantic. *Canadian Journal of Fisheries and Aquatic Sciences* 49 (7): 1474–1477.

Conrath, C. L., and J. A. Musick. 2007. The sandbar shark summer nursery within bays and lagoons of the eastern shore of Virginia. *Transactions of the American Fisheries Society* 136 (4): 999–1007.

Daly-Engel, T. S., R. D. Grubbs, B. W. Bowen, and R. J. Toonen. 2007. Frequency of multiple paternity in an unexploited tropical population of sandbar sharks (*Carcharhinus plumbeus*). *Canadian Journal of Fisheries and Aquatic Sciences* 64 (2): 198–204.

Diatta, Y., A. S. I. Amadou, C. Reynaud, O. Guelorget, and C. Capape. 2008. New biological observations on the sandbar shark *Carcharhinus plumbeus* (Chondrichthyes: Carcharhinidae) from the coast of Senegal (eastern tropical Atlantic). *Cahiers de Biologie Marine* 49 (2): 103–111.

Dowd, W. W., R. W. Brill, P. G. Bushnell, and J. A. Musick. 2006. Standard and routine metabolic rates of juvenile sandbar sharks (*Carcharhinus plumbeus*), including the effects of body mass and acute temperature change. *Fishery Bulletin* 104 (3): 323–331.

Joung, S. J., and C. T. Chen. 1995. Reproduction in the sandbar shark, *Carcharhinus plumbeus,* in the waters off northeastern Taiwan. *Copeia* 1995:659–665.

Joung, S. J., Y. Y. Liao, and C. T. Chen. 2004. Age and growth of sandbar shark, *Carcharhinus plumbeus,* in northeastern Taiwan waters. *Fisheries Research* 70 (1): 83–96.

Marshall, H., G. Skomal, P. G. Ross, and D. Bernal. 2015. At-vessel and post-release mortality of the dusky (Carcharhinus obscurus) and sandbar (C. plumbeus) sharks after longline capture. *Fisheries Research* 172:373–384.

McAuley, R. B., C. A. Simpfendorfer, G. A. Hyndes, R. R. Allison, J. A. Chidlow, S. J. Newman, and R. C. J. Lenanton. 2006. Validated age and growth of the sandbar shark, *Carcharhinus plumbeus* (Nardo 1827) in the waters off Western Australia. *Environmental Biology of Fishes* 77 (3–4): 385–400.

McAuley, R. B., C. A. Simpfendorfer, G. A. Hyndes, and R. C. J. Lenanton. 2007. Distribution and reproductive biology of the sandbar shark, *Carcharhinus plumbeus* (Nardo), in Western Australian waters. *Marine and Freshwater Research* 58 (1): 116–126.

McElroy, W. D., B. M. Wetherbee, C. S. Mostello, C. G. Lowe, G. L. Crow, and R. C. Wass. 2006. Food habits and ontogenetic changes in the diet of the sandbar shark, *Carcharhinus plumbeus,* in Hawaii. *Environmental Biology of Fishes* 76 (1): 81–92.

Merson, R. R., and H. L. Pratt. 2001. Distribution, movements and growth of young sandbar sharks, *Carcharhinus plumbeus,* in the nursery grounds of Delaware Bay. *Environmental Biology of Fishes* 61 (1): 13–24.

Portnoy, D. S., A. N. Piercy, J. A. Musick, G. H. Burgess, and J. E. Graves. 2007. Genetic polyandry and sexual conflict in the sandbar shark, *Carcharhinus plumbeus,* in the western North Atlantic and Gulf of Mexico. *Molecular Ecology* 16 (1): 187–197.

Romine, J. G., R. D. Grubbs, and J. A. Musick. 2006. Age and growth of the sandbar shark, *Carcharhinus plumbeus,* in Hawaiian waters through vertebral analysis. *Environmental Biology of Fishes* 77 (3–4): 229–239.

Romine, J. G., J. A. Musick, and R. A. Johnson. 2013. Compensatory growth of the sandbar shark in the western North Atlantic including the Gulf of Mexico. *Marine and Coastal Fisheries* 5 (1): 189–199.

Saidi, B., M. N. Bradai, A. Bouain, and C. Capape. 2007. Feeding habits of the sandbar shark *Carcharhinus plumbeus* (Chondrichthyes: Carcharhinidae) from the Gulf of Gabes, Tunisia. *Cahiers de Biologie Marine* 48 (2): 139–144.

Shiffman, D. S., B. S. Frazier, J. R. Kucklick, D. Abel, J. Brandes, and G. Sancho. 2014. Feeding ecology of the sandbar shark in South Carolina estuaries revealed through delta C-13 and delta N-15 stable isotope analysis. *Marine and Coastal Fisheries* 6 (1): 156–169.

Sandtiger Shark

Bansemer, C. S., and M. B. Bennett. 2011. Sex and maturity-based differences in movement and migration patterns of grey nurse shark, *Carcharias taurus,* along the eastern coast of Australia. *Marine and Freshwater Research* 62 (6): 596–606.

Branstetter, S., and J. A. Musick. 1994. Age and growth estimates for the sand tiger in the northwestern Atlantic Ocean. *Transactions of the American Fisheries Society* 123 (2): 242–254.

Chapman, D. D., S. P. Wintner, D. L. Abercrombie, J. Ashe, A. M. Bernard, M. S. Shivji, and K. A. Feldheim. 2013. The behavioural and genetic mating system of the sand tiger shark, *Carcharias taurus,* an intrauterine cannibal. *Biology Letters* 9 (3). doi.org/10.1098/rsbl.2013.0003.

Dicken, M. L., A. J. Booth, and M. J. Smale. 2008. Estimates of juvenile and adult raggedtooth shark (*Carcharias taurus*) abundance along the east coast of South Africa. *Canadian Journal of Fisheries and Aquatic Sciences* 65 (4): 621–632.

Dicken, M. L., A. J. Booth, M. J. Smale, and G. Cliff. 2007. Spatial and seasonal distribution patterns of juvenile and adult raggedtooth sharks (*Carcharias taurus*) tagged off the east coast of South Africa. *Marine and Freshwater Research* 58 (1): 127–134.

Gilmore, R. G. 1993. Reproductive biology of lamnoid sharks. *Environmental Biology of Fishes* 38:95–114.

Gilmore, R. G., J. W. Dodrill, and P. A. Linley. 1983. Reproduction and embryonic development of the sand tiger shark, *Odontaspis taurus* (Rafinesque). *Fishery Bulletin* 81:201–225.

Goldman, K. J., S. Branstetter, and J. A. Musick. 2006. A re-examination of the age and growth of sand tiger sharks, *Carcharias taurus,* in the western North Atlantic: The importance of ageing protocols and use of multiple back-calculation techniques. *Environmental Biology of Fishes* 77 (3–4): 241–252.

Hamlett, W. C., and M. Hysell. 1998. Uterine specializations in elasmobranchs. *Journal of Experimental Zoology* 282 (4–5): 438–459.

Kneebone, J., J. Chisholm, D. Bernal, and G. Skomal. 2013. The physiological effects of capture stress, recovery, and post-release survivorship of juvenile sand tigers (*Carcharias taurus*) caught on rod and reel. *Fisheries Research* 147:103–114.

Lucifora, L. O., R. C. Menni, and A. H. Escalante. 2002. Reproductive ecology and abundance of the sand tiger shark, *Carcharias taurus,* from the southwestern Atlantic. *Ices Journal of Marine Science* 59 (3): 553–561.

Nichols, J. T., and R. C. Murphy. 1916. Long Island fauna—IV: The sharks. *Brooklyn Museum Science Bulletin* 3 (1): 21–22.

Otway, N. M., and M. T. Ellis. 2011. Pop-up archival satellite tagging of *Carcharias taurus:* Movements and depth/temperature-related use of southeastern Australian waters. *Marine and Freshwater Research* 62 (6): 607–620.

Passerotti, M. S., A. H. Andrews, J. K. Carlson, S. P. Wintner, K. J. Goldman, and L. J. Natanson. 2014. Maximum age and missing time in the vertebrae of sand tiger shark (*Carcharias taurus*): validated lifespan from bomb radiocarbon dating in the western North Atlantic and southwestern Indian Oceans. *Marine and Freshwater Research* 65 (8): 674–687.

Smale, M. J. 2002. Occurrence of *Carcharias taurus* in nursery areas of the Eastern and Western Cape, South Africa. *Marine & Freshwater Research* 53 (2): 551–556.

———. 2005. The diet of the ragged-tooth shark *Carcharias taurus* Rafinesque 1810 in the Eastern Cape, South Africa. *African Journal of Marine Science* 27 (1): 331–335.

Teter, S. M., B. M. Wetherbee, D. A. Fox, C. H. Lam, D. A. Kiefer, and M. Shivji. 2015. Migratory patterns and habitat use of the sand tiger shark (Carcharias taurus) in the western North Atlantic. *Marine and Freshwater Research* 66 (2): 158–169.

Silky Shark

Alejo-Plata, C., J. L. Gomez-Marquez, S. Ramos, and E. Herrera. 2007. Presence of neonates and juvenile scalloped hammerhead sharks *Sphyrna lewini* (Griffith & Smith, 1834) and silky sharks *Carcharhinus falciformis* (Muller & Henle, 1839) in the Oaxaca coast, Mexico. *Revista de Biologia Marina y Oceanografia* 42 (3): 403–413.

Clarke, C., J. S. E. Lea, and R. F. G. Ormond. 2011. Reef-use and residency patterns of a baited population of silky sharks, *Carcharhinus falciformis,* in the Red Sea. *Marine and Freshwater Research* 62 (6): 668–675.

———. 2013. Changing relative abundance and behaviour of silky and grey reef sharks baited over 12 years on a Red Sea reef. *Marine and Freshwater Research* 64 (10): 909–919.

Duffy, L. M., R. J. Olson, C. E. Lennert-Cody, F. Galvan-Magana, N. Bocanegra-Castillo, and P. M. Kuhnert. 2015. Foraging ecology of silky sharks,

Carcharhinus falciformis, captured by the tuna purse-seine fishery in the eastern Pacific Ocean. *Marine Biology* 162 (3): 571–593.

Filmalter, J. D., L. Dagorn, P. D. Cowley, and M. Taquet. 2011. First descriptions of the behavior of silky sharks, *Carcharhinus falciformis,* around drifting fish aggregating devices in the Indian Ocean. *Bulletin of Marine Science* 87 (3): 325–337.

Galvan-Tirado, C., F. Galvan-Magana, and R. I. Ochoa-Baez. 2015. Reproductive biology of the silky shark *Carcharhinus falciformis* in the southern Mexican Pacific. *Journal of the Marine Biological Association of the United Kingdom* 95 (3): 561–567.

Hall, N. G., C. Bartron, W. T. White, Dharmadi, and I. C. Potter. 2012. Biology of the silky shark *Carcharhinus falciformis* (Carcharhinidae) in the eastern Indian Ocean, including an approach to estimating age when timing of parturition is not well defined. *Journal of Fish Biology* 80 (5): 1320–1341.

Joung, S. J., C. T. Chen, H. H. Lee, and K. M. Liu. 2008. Age, growth, and reproduction of silky sharks, *Carcharhinus falciformis,* in northeastern Taiwan waters. *Fisheries Research* 90 (1–3): 78–85.

Oshitani, S., S. Nakano, and S. Tanaka. 2003. Age and growth of the silky shark *Carcharhinus falciformis* from the Pacific Ocean. *Fisheries Science* 69 (3): 456–464.

Poisson, F., J. D. Filmalter, A. L. Vernet, and L. Dagorn. 2014. Mortality rate of silky sharks (*Carcharhinus falciformis*) caught in the tropical tuna purse seine fishery in the Indian Ocean. *Canadian Journal of Fisheries and Aquatic Sciences* 71 (6): 795–798.

Rego, M. G., F. H. V. Hazin, J. E. Neto, P. G. V. Oliveira, M. G. Soares, K. R. L. S. Torres, F. O. Lana, P. C. G. Roque, N. L. Santos, and R. Coelho. 2013. Morphological analysis and description of the ovaries of female silky sharks, *Carcharhinus falciformis* (Muller & Henle, 1839). *Neotropical Ichthyology* 11 (4): 815–819.

Sanchez–de Ita, J. A., C. Quinonez-Velazquez, F. Galvan-Magana, N. Bocanegra-Castillo, and R. Felix-Uraga. 2011. Age and growth of the silky shark *Carcharhinus falciformis* from the west coast of Baja California Sur, Mexico. *Journal of Applied Ichthyology* 27 (1): 20–24.

Santana-Hernandez, H., J. Tovar-Avila, and J. J. Valdez-Flores. 2014. Estimation of the total, fork and precaudal lengths for the silky shark, *Carcharhinus falciformis* (Carcharhiniformes: Carcharhinidae), from the interdorsal length. *Hidrobiologica* 24 (2): 159–162.

Smoothhound Sharks

Carlson, J. K., and G. R. Parsons. 2001. The effects of hypoxia on three sympatric shark species: Physiological and behavioral responses. *Environmental Biology of Fishes* 61 (4): 427–433.

Conrath, C. L., J. Gelsleichter, and J. A. Musick. 2002. Age and growth of the smooth dogfish (*Mustelus canis*) in the northwest Atlantic Ocean. *Fishery Bulletin* 100 (4): 674–682.

Gelsleichter, J., J. A. Musick, and S. Nichols. 1999. Food habits of the smooth dogfish, *Mustelus canis,* dusky shark, *Carcharhinus obscurus,* Atlantic sharpnose shark, *Rhizoprionodon terraenovae,* and the sand tiger, *Carcharias taurus,* from the Northwest Atlantic Ocean. *Environmental Biology of Fishes* 54 (2): 205–217.

Giresi, M., M. A. Renshaw, D. S. Portnoy, and J. R. Gold. 2012. Isolation and characterization of microsatellite markers for the dusky smoothhound shark, *Mustelus canis. Conservation Genetics Resources* 4 (1): 101–104.

Hamlett, W. C., J. A. Musick, C. K. Hysell, and D. M. Sever. 2002. Uterine epithelial-sperm interaction, endometrial cycle and sperm storage in the terminal zone of the oviducal gland in the placental smoothhound, *Mustelus canis. Journal of Experimental Zoology* 292 (2): 129–144.

Kalinoski, M., A. Hirons, A. Horodysky, and R. Brill. 2014. Spectral sensitivity, luminous sensitivity, and temporal resolution of the visual systems in three sympatric temperate coastal shark species. *Journal of Comparative Physiology A—Neuroethology Sensory Neural and Behavioral Physiology* 200 (12): 997–1013.

Tavares, R., M. Lemus, and K. S. Chung. 2006. Evaluation of the instantaneous growth of juvenile smooth dogfish sharks (*Mustelus canis*) in their natural habitat, based on the RNA/DNA ratio. *Ciencias Marinas* 32 (2): 297–302.

Vooren, C. M. 1992. Reproductive strategies of 8 species of viviparous elasmobranchs from Southern Brazil. *Bulletin de la Société Zoologique de France—Evolution et Zoologie* 117 (3): 303–312.

Zagaglia, C. R., C. Damiano, F. H. V. Hazin, and M. K. Broadhurst. 2011. Reproduction in *Mustelus canis* (Chondrichthyes: Triakidae) from an unexploited population off northern Brazil. *Journal of Applied Ichthyology* 27 (1): 25–29.

Spinner Shark

Allen, B. R., and S. P. Wintner. 2002. Age and growth of the spinner shark *Carcharhinus brevipinna* (Muller and Henle, 1839) off the Kwazulu-Natal Coast, South Africa. *South African Journal of Marine Science* 24:1–8.

Bethea, D. M., J. A. Buckel, and J. K. Carlson. 2004. Foraging ecology of the early life stages of four sympatric shark species. *Marine Ecology—Progress Series* 268:245–264.

Carlson, J. K., and I. E. Baremore. 2005. Growth dynamics of the spinner shark (*Carcharhinus brevipinna*) off the United States southeast and Gulf of Mexico coasts: A comparison of methods. *Fishery Bulletin* 103 (2): 280–291.

Carlson, J. K., L. F. Hale, A. Morgan, and G. Burgess. 2012. Relative abundance and size of coastal sharks derived from commercial shark longline catch and effort data. *Journal of Fish Biology* 80 (5): 1749–1764.

Geraghty, P. T., J. E. Williamson, W. G. Macbeth, S. P. Wintner, A. V. Harry, J. R. Ovenden, and M. R. Gillings. 2013. Population expansion and genetic structure in *Carcharhinus brevipinna* in the Southern Indo-Pacific. *PLoS One* 8 (9): e75169. doi:10.1371/journal.pone.0075169.

Heist, E. J., and J. R. Gold. 1999. Genetic identification of sharks in the US Atlantic large coastal shark fishery. *Fishery Bulletin* 97 (1): 53–61.

Hussey, N. E., S. P. Wintner, S. F. J. Dudley, G. Cliff, D. T. Cocks, and M. A. MacNeil. 2010. Maternal investment and size-specific reproductive output in Carcharhinid sharks. *Journal of Animal Ecology* 79 (1): 184–193.

Joung, S. J., Y. Y. Liao, K. M. Liu, C. T. Chen, and L. C. Leu. 2005. Age, growth, and reproduction of the spinner shark, *Carcharhinus brevipinna,* in the northeastern waters of Taiwan. *Zoological Studies* 44 (1): 102–110.

Taylor, S., W. Sumpton, and T. Ham. 2011. Fine-scale spatial and seasonal partitioning among large sharks and other elasmobranchs in south-eastern Queensland, Australia. *Marine and Freshwater Research* 62 (6): 638–647.

Tiger Shark

Afonso, A. S., F. H. V. Hazin, R. R. Barreto, F. M. Santana, and R. P. Lessa. 2012. Extraordinary growth in tiger sharks *Galeocerdo cuvier* from the South Atlantic Ocean. *Journal of Fish Biology* 81 (6): 2080–2085.

Driggers, W. B., G. W. Ingram, M. A. Grace, C. T. Gledhill, T. A. Henwood, C. N. Horton, and C. M. Jones. 2008. Pupping areas and mortality rates of young tiger sharks *Galeocerdo cuvier* in the western North Atlantic Ocean. *Aquatic Biology* 2 (2): 161–170.

Heithaus, M. R., L. M. Dill, G. J. Marshall, and B. Buhleier. 2002. Habitat use and foraging behavior of tiger sharks (*Galeocerdo cuvier*) in a seagrass ecosystem. *Marine Biology* 140 (2): 237–248.

Holland, K. N., B. M. Wetherbee, C. G. Lowe, and C. G. Meyer. 1999. Movements of tiger sharks (*Galeocerdo cuvier*) in coastal Hawaiian waters. *Marine Biology (Berlin)* 134 (4): 665–673.

Kneebone, J., L. J. Natanson, A. H. Andrews, and W. H. Howell. 2008. Using bomb radiocarbon analyses to validate age and growth estimates for the tiger shark, *Galeocerdo cuvier,* in the western North Atlantic. *Marine Biology* 154 (3): 423–434.

Natanson, L. J., J. G. Casey, N. E. Kohler, and T. Colket. 1999. Growth of the tiger shark, *Galeocerdo cuvier,* in the western North Atlantic based on tag returns and length frequencies; and a note on the effects of tagging. *Fishery Bulletin* 97 (4): 944–953.

Simpfendorfer, Colin A., Adrian B. Goodreid, and Rory B. McAuley. 2001. Size, sex and geographic varia-tion in the diet of the tiger shark, *Galeocerdo cuvier,* from Western Australian waters. *Environmental Biology of Fishes* 61 (1): 37–46.

Whitney, N. M., and G. L. Crow. 2007. Reproductive biology of the tiger shark (*Galeocerdo cuvier*) in Hawaii. *Marine Biology* 151 (1): 63–70.

Whale Shark

Berumen, M. L., C. D. Braun, J. E. M. Cochran, G. B. Skomal, and S. R. Thorrold. 2014. Movement patterns of juvenile whale sharks tagged at an aggregation site in the Red Sea. *PLoS One* 9 (7): e103536. doi:10.1371/journal.pone.0103536.

Brunnschweiler, J. M., H. Baensch, S. J. Pierce, and D. W. Sims. 2009. Deep-diving behaviour of a whale shark *Rhincodon typus* during long-distance movement in the western Indian Ocean. *Journal of Fish Biology* 74 (3): 706–714.

Burks, C. M., W. B. Driggers, and K. D. Mullin. 2006. Abundance and distribution of whale sharks (*Rhincodon typus*) in the northern Gulf of Mexico. *Fishery Bulletin* 104 (4): 579–584.

De la Parra Venegas, R., R. Hueter, J. González Cano, J. Tyminski, J. Gregorio Remolina, M. Maslanka, A. Ormos, L. Weigt, B. Carlson, and A. Dove. 2011. An unprecedented aggregation of whale sharks, *Rhincodon typus,* in Mexican coastal waters of the Caribbean Sea. *PLoS One* 6 (4): e18994. doi:10.1371/journal.pone.0018994.

Duffy, C. A. J. 2002. Distribution, seasonality, lengths, and feeding behaviour of whale sharks (*Rhincodon typus*) observed in New Zealand waters. *New Zealand Journal of Marine and Freshwater Research* 36 (3): 565–570.

Eckert, S. A., L. L. Dolar, G. L. Kooyman, W. F. Perrin, and R. A. Rahman. 2002. Movements of whale sharks (*Rhincodon typus*) in South-east Asian waters as determined by satellite telemetry. *Journal of Zoology (London)* 257 (1): 111–115.

Fox, S., I. Foisy, R. D. Venegas, B. E. G. Pastoriza, R. T. Graham, E. R. Hoffmayer, J. Holmberg, and S. J. Pierce. 2013. Population structure and residency of whale sharks *Rhincodon typus* at Utila, Bay Islands, Honduras. *Journal of Fish Biology* 83 (3): 574–587.

Gifford, A., L. J. V. Compagno, M. Levine, and A. Antoniou. 2007. Satellite tracking of whale sharks using tethered tags. *Fisheries Research* 84 (1): 17–24.

Graham, R. T., and C. M. Roberts. 2007. Assessing the size, growth rate and structure of a seasonal population of whale sharks (*Rhincodon typus* Smith 1828) using conventional tagging and photo identification. *Fisheries Research* 84 (1): 71–80.

Gunn, J. S., J. D. Stevens, T. L. O. Davis, and B. M. Norman. 1999. Observations on the short-term movements and behaviour of whale sharks (*Rhincodon typus*) at Ningaloo Reef, Western Australia. *Marine Biology (Berlin)* 135 (3): 553–559.

Hacohen-Domene, A., R. O. Martinez-Rincon, F. Galvan-Magana, N. Cardenas-Palomo, R. de la

Parra–Venegas, B. Galvan-Pastoriza, and Alistair D. M. Dove. 2015. Habitat suitability and environmental factors affecting whale shark (*Rhincodon typus*) aggregations in the Mexican Caribbean. *Environmental Biology of Fishes* 98 (8): 1953–1964.

Haskell, P. J., A. McGowan, A. Westling, A. Mendez-Jimenez, C. A. Rohner, K. Collins, M. Rosero-Caicedo, J. Salmond, A. Monadjem, A. D. Marshall, and Simon J. Pierce. 2015. Monitoring the effects of tourism on whale shark *Rhincodon typus* behaviour in Mozambique. *Oryx* 49 (3): 492–499.

Hsu, H. H., S. J. Joung, R. E. Hueter, and K. M. Liu. 2014. Age and growth of the whale shark (*Rhincodon typus*) in the north-western Pacific. *Marine and Freshwater Research* 65 (12): 1145–1154.

Hsu, H. H., S. J. Joung, Y. Y. Liao, and K. M. Liu. 2007. Satellite tracking of juvenile whale sharks, *Rhincodon typus,* in the Northwestern Pacific. *Fisheries Research* 84 (1): 25–31.

Hsu, H. H., C. Y. Lin, and S. J. Joung. 2014. The first record, tagging and release of a neonatal whale shark *Rhincodon typus* in Taiwan. *Journal of Fish Biology* 85 (5): 1753–1756.

Hueter, R. E., J. P. Tyminski, and R. de la Parra. 2013. Horizontal movements, migration patterns, and population structure of whale sharks in the Gulf of Mexico and Northwestern Caribbean Sea. *PLoS One* 8 (8): e71883. doi:10.1371/journal.pone.0071883.

Meekan, M. G., S. N. Jarman, C. McLean, and M. B. Schultz. 2009. DNA evidence of whale sharks (*Rhincodon typus*) feeding on red crab (Gecarcoidea natalis) larvae at Christmas Island, Australia. *Marine and Freshwater Research* 60 (6): 607–609.

Motta, P. J., M. Maslanka, R. E. Hueter, R. L. Davis, R. de la Parra, S. L. Mulvany, M. L. Habegger, et al. 2010. Feeding anatomy, filter-feeding rate, and diet of whale sharks *Rhincodon typus* during surface ram filter feeding off the Yucatan Peninsula, Mexico. *Zoology* 113 (4): 199–212.

Rohner, C. A., A. J. Richardson, A. D. Marshall, S. J. Weeks, and S. J. Pierce. 2011. How large is the world's largest fish? Measuring whale sharks *Rhincodon typus* with laser photogrammetry. *Journal of Fish Biology* 78 (1): 378–385.

Rowat, D., and K. S. Brooks. 2012. A review of the biology, fisheries and conservation of the whale shark *Rhincodon typus*. *Journal of Fish Biology* 80 (5): 1019–1056.

Taylor, J. G. 2007. Ram filter-feeding and nocturnal feeding of whale sharks (*Rhincodon typus*) at Ningaloo Reef, Western Australia. *Fisheries Research* 84 (1): 65–70.

Tyminski, J., R. de la Parra–Venegas, J. González Cano, and R. Hueter. 2015. Vertical movements and patterns in diving behavior of whale sharks as revealed by pop-up satellite tags in the eastern Gulf of Mexico. *PLoS One* 10 (11): e0142156. doi:10.1371/journal.pone.0142156.

Wilson, S. G., J. G. Taylor, and A. F. Pearce. 2001. The seasonal aggregation of whale sharks at Ningaloo Reef, Western Australia: Currents, migrations and the El Nino/Southern Oscillation. *Environmental Biology of Fishes* 61 (1): 1–11.

Wintner, S. P. 2000. Preliminary study of vertebral growth rings in the whale shark, *Rhincodon typus,* from the east coast of South Africa. *Environmental Biology of Fishes* 59 (4): 441–451.

Skates and Rays, Including Sawfish
Atlantic Stingray and Southern Stingray

Enzor, L. A., R. E. Wilborn, and W. A. Bennett. 2011. Toxicity and metabolic costs of the Atlantic stingray (*Dasyatis sabina*) venom delivery system in relation to its role in life history. *Journal of Experimental Marine Biology and Ecology* 409 (1–2): 235–239.

Grim, J. M., A. A. Ding, and W. A. Bennett. 2012. Differences in activity level between cownose rays (*Rhinoptera bonasus*) and Atlantic stingrays (*Dasyatis sabina*) are related to differences in heart mass, hemoglobin concentration, and gill surface area. *Fish Physiology and Biochemistry* 38 (5): 1409–1417.

Hamlett, W. C., J. A. Musick, A. M. Eulitt, R. L. Jarrell, and M. A. Kelly. 1996. Ultrastructure of uterine trophonemata, accommodation for uterolactation, and gas exchange in the southern stingray, *Dasyatis americana*. *Canadian Journal of Zoology* 74 (8): 1417–1430.

Henningsen, A. D. 2000. Notes on reproduction in the southern stingray, *Dasyatis americana* (Chondrichthyes: Dasyatidae), in a captive environment. *Copeia* 3:826–828.

Johnson, M. R., and F. F. Snelson. 1996. Reproductive life history of the Atlantic stingray, *Dasyatis sabina* (Pisces, Dasyatidae), in the freshwater St. Johns River, Florida. *Bulletin of Marine Science* 59 (1): 74–88.

Kajiura, S. M., A. P. Sebastian, and T. C. Tricas. 2000. Dermal bite wounds as indicators of reproductive seasonality and behaviour in the Atlantic stingray, *Dasyatis sabina*. *Environmental Biology of Fishes* 58 (1): 23–31.

Kajiura, S. M., and T. C. Tricas. 1996. Seasonal dynamics of dental sexual dimorphism in the Atlantic stingray *Dasyatis sabina*. *Journal of Experimental Biology* 199 (10): 2297–2306.

McGowan, D. W., and S. M. Kajiura. 2009. Electroreception in the euryhaline stingray, *Dasyatis sabina*. *Journal of Experimental Biology* 212 (10): 1544–1552.

Piercy, A., J. Gelsleichter, and F. F. Snelson. 2006. Morphological changes in the clasper gland of the Atlantic stingray, *Dasyatis sabina,* associated with the seasonal reproductive cycle. *Journal of Morphology* 267 (1): 109–114.

Piermarini, P. M., and D. H. Evans. 1998. Osmoregulation of the Atlantic stingray (*Dasyatis sabina*) from the freshwater lake Jesup of the St. Johns River, Florida. *Physiological Zoology* 71 (5): 553–560.

Tilley, A., J. Lopez-Angarita, and J. R. Turner. 2013. Effects of scale and habitat distribution on the movement of the southern stingray *Dasyatis americana* on a Caribbean atoll. *Marine Ecology Progress Series* 482:169–179.

Wallman, H. L., and W. A. Bennett. 2006. Effects of parturition and feeding on thermal preference of Atlantic stingray, *Dasyatis sabina* (Lesueur). *Environmental Biology of Fishes* 75 (3): 259–267.

Clearnose Skate

Hamlett, W. C., and M. Hysell. 1998. Uterine specializations in elasmobranchs. *Journal of Experimental Zoology* 282 (4–5): 438–459.

Luer, C. A., C. J. Walsh, A. B. Bodine, and J. T. Wyffels. 2007. Normal embryonic development in the clearnose skate, *Raja eglanteria,* with experimental observations on artificial insemination. *Environmental Biology of Fishes* 80 (2–3): 239–255.

Macesic, L. J., and S. M. Kajiura. 2010. Comparative punting kinematics and pelvic fin musculature of benthic batoids. *Journal of Morphology* 271 (10): 1219–1228.

Cownose Ray

Ajemian, M. J., and S. P. Powers. 2012. Habitat-specific feeding by cownose rays (*Rhinoptera bonasus*) of the northern Gulf of Mexico. *Environmental Biology of Fishes* 95 (1): 79–97.

Collins, A. B., M. R. Heupel, and P. J. Motta. 2007. Residence and movement patterns of cownose rays *Rhinoptera bonasus* within a south-west Florida estuary. *Journal of Fish Biology* 71 (4): 1159–1178.

Sasko, D. E., M. N. Dean, P. J. Motta, and R. E. Hueter. 2006. Prey capture behavior and kinematics of the Atlantic cownose ray, *Rhinoptera bonasus*. *Zoology* 109 (3): 171–181.

Mantas and Mobulid Rays

Braun, C. D., G. B. Skomal, S. R. Thorrold, and M. L. Berumen. 2014. Diving behavior of the reef manta ray links coral reefs with adjacent deep pelagic habitats. *PLoS One* 9 (2): e88170. doi:10.1371/journal.pone.0088170.

Couturier, L. I. E., A. D. Marshall, F. R. A. Jaine, K. Kashiwagi, S. J. Pierce, K. A Townsend, S. J. Weeks, M. B. Bennett, and A. J. Richardson. 2012. Biology, ecology, and conservation of the Mobulidae. *Journal of Fish Biology* 80:1075–1119.

Deakos, M. H. 2012. The reproductive ecology of resident manta rays (*Manta alfredi*) off Maui, Hawaii, with an emphasis on body size. *Environmental Biology of Fishes* 94 (2): 443–456.

Di Sciara, G. N., G. Lauriano, N. Pierantonio, A. Canadas, G. Donovan, and S. Panigada. 2015. The devil we don't know: Investigating habitat and abundance of endangered giant devil rays in the northwestern Mediterranean Sea. *PLoS One* 10 (11): e0141189. doi:10.1371/journal. pone.0141189.

Luiz, O. J., A. P. Balboni, G. Kodja, M. Andrade, and H. Marum. 2009. Seasonal occurrences of *Manta birostris* (Chondrichthyes: Mobulidae) in southeastern Brazil. *Ichthyological Research* 56 (1): 96–99.

O'Shea, O. R., M. J. Kingsford, and J. Seymour. 2010. Tide-related periodicity of manta rays and sharks to cleaning stations on a coral reef. *Marine and Freshwater Research* 61 (1): 65–73.

Smith, W. D., J. J. Bizzarro, and G. M. Cailliet. 2009. The artisanal elasmobranch fishery on the east coast of Baja California, Mexico: Characteristics and management considerations. *Ciencias Marinas* 35 (2): 209–236.

Roughtail Stingray

Capapé, C. 1993. New data on the reproductive biology of the thorny stingray, *Dasyatis centroura* (Pisces, Dasyatidae) from off the Tunisian coasts. *Environmental Biology of Fishes* 38 (1–3): 73–80.

Faria, V. V., L. S. Rolim, L. A. L. Vaz, and M. A. A. Furtado-Neto. 2012. Reevaluation of RAPD markers involved in a case of stingray misidentification (Dasyatidae: Dasyatis). *Genetics and Molecular Research* 11 (4): 3835–3845.

Schmidt, B. F., A. F. Amorim, and A. W. S. Hilsdorf. 2015. PCR-RFLP analysis to identify four ray species of the genus Dasyatis (Elasmobranchii, Dasyatidae) fished along the southeastern and southern coast of Brazil. *Fisheries Research* 167:71–74.

Zogaris, S., and A. De Maddalena. 2014. Sharks, blast fishing and shifting baselines: Insights from Hass's 1942 Aegean expedition. *Cahiers de Biologie Marine* 55 (3): 305–313.

Sawfish

Carlson, J. K., S. J. B. Gulak, C. A. Simpfendorfer, R. D. Grubbs, J. G. Romine, and G. H. Burgess. 2014. Movement patterns and habitat use of smalltooth sawfish, *Pristis pectinata,* determined using pop-up satellite archival tags. *Aquatic Conservation— Marine and Freshwater Ecosystems* 24 (1): 104–117.

Carlson, J. K., and C. A. Simpfendorfer. 2015. Recovery potential of smalltooth sawfish, *Pristis pectinata,* in the United States determined using population viability models. *Aquatic Conservation—Marine and Freshwater Ecosystems* 25 (2): 187–200.

Faria, V. V., M. T. McDavitt, P. Charvet, T. R. Wiley, C. A. Simpfendorfer, and G. J. P. Naylor. 2013. Species delineation and global population structure of critically endangered sawfishes (Pristidae). *Zoological Journal of the Linnean Society* 167 (1): 136–164.

Guttridge, T. L., S. J. B. Gulak, B. R. Franks, J. K. Carlson, S. H. Gruber, K. S. Gledhill, M. E. Bond, G. Johnson, and R. D. Grubbs. 2015. Occurrence and habitat use of the critically endangered smalltooth sawfish *Pristis pectinata* in the Bahamas. *Journal of Fish Biology* 87 (6): 1322–1341.

Poulakis, G. R., P. W. Stevens, A. A. Timmers, T. R. Wiley, and C. A. Simpfendorfer. 2011. Abiotic affinities and spatiotemporal distribution of the endangered smalltooth sawfish, *Pristis pectinata,* in a southwestern Florida nursery. *Marine and Freshwater Research* 62 (10): 1165–1177.

Scharer, R. M., W. F. Patterson, J. K. Carlson, and G. R. Poulakis. 2012. Age and growth of endangered smalltooth sawfish (*Pristis pectinata*) verified with LA-ICP-MS analysis of vertebrae. *PLoS One* 7 (10): e47850. doi:10.1371/journal.pone.0047850.

Seitz, J. C., and G. R. Poulakis. 2006. Anthropogenic effects on the smalltooth sawfish (*Pristis pectinata*) in the United States. *Marine Pollution Bulletin* 52 (11): 1533–1540.

Simpfendorfer, C. A., G. R. Poulakis, P. M. O'Donnell, and T. R. Wiley. 2008. Growth rates of juvenile smalltooth sawfish *Pristis pectinata* Latham in the western Atlantic. *Journal of Fish Biology* 72 (3): 711–723.

Simpfendorfer, C. A., T. R. Wiley, and B. G. Yeiser. 2010. Improving conservation planning for an endangered sawfish using data from acoustic telemetry. *Biological Conservation* 143 (6): 1460–1469.

Simpfendorfer, C. A., B. G. Yeiser, T. R. Wiley, G. R. Poulakis, P. W. Stevens, and M. R. Heupel. 2011. Environmental influences on the spatial ecology of juvenile smalltooth sawfish (*Pristis pectinata*): Results from acoustic monitoring. *PLoS One* 6 (2): e16918. doi:10.1371/journal.pone.0016918.

Spotted Eagle Ray

Bassos-Hull, K., K. A. Wilkinson, P. T. Hull, D. A. Dougherty, K. L. Omori, L. E. Ailloud, J. J. Morris, and R. E. Hueter. 2014. Life history and seasonal occurrence of the spotted eagle ray, *Aetobatus narinari,* in the eastern Gulf of Mexico. *Environmental Biology of Fishes* 97 (9): 1039–1056.

Newby, J., T. Darden, K. Bassos-Hull, and A. M. Shedlock. 2014. Kin structure and social organization in the spotted eagle ray, *Aetobatus narinari,* off coastal Sarasota, FL. *Environmental Biology of Fishes* 97 (9): 1057–1065.

Sellas, A. B., K. Bassos-Hull, J. Carlos Perez-Jimenez, J. Alberto Angulo-Valdes, M. A. Bernal, and R. E. Hueter. 2015. Population structure and seasonal migration of the spotted eagle ray, *Aetobatus narinari. Journal of Heredity* 106 (3): 266–275.

Yellow Stingray

Spieler, R. E., D. P. Fahy, R. L. Sherman, J. A. Sulikowski, and T. P. Quinn. 2013. The yellow stingray, *Urobatis jamaicensis* (Chondrichthyes: Urotrygonidae): A synoptic review. *Caribbean Journal of Science* 47 (1): 67–97.

Index

The Author

Dr. Jeffrey Carrier is a Professor Emeritus of Biology at Albion College where he taught for more than thirty years. He grew up along the northeast coast of Florida, where fishing and diving dominated his free time and introduced him to the oceanic world and the sharks that were found there. During his career as a college professor, he was the recipient of numerous teaching and research awards. He has published many articles in both the scientific literature and in the popular press, including *National Geographic* Magazine. Carrier's research has been featured in many television shows for the National Geographic Society and the Discovery Channel's *Shark Week*. Dr. Carrier has written and edited four books detailing the biology of sharks and their relatives, including one book written for children. His research has been centered in the Florida Keys, where he continues to study nurse shark aging, growth, migration, and reproductive biology. Active in the American Elasmobranch Society, an international organization of scientists, educators, conservationists, and resource managers, he has been elected twice as that professional organization's president.

The Photographers

Andy Murch is an award-winning wildlife photographer with an obsession for photographing sharks and rays. Born with an insatiable thirst for adventure, Andy escaped from the United Kingdom in the 1980s and wandered the planet collecting experiences and images. After working as a diving instructor and as a submersible pilot, he became a full-time photographer in 2003. His images have since appeared in countless books and magazines around the world. Andy is the creator of the ever-expanding shark information database on Elasmodiver.com. He is also the driving force behind the Predators in Peril Project, an initiative that captures and freely shares images of endangered sharks and draws attention to the plight of many lesser-known endangered species that desperately need some time in the limelight. You can follow Andy's conservation work at predatorsinperil.org. Between photographic assignments and conservation field trips, Andy runs Big Fish Expeditions, a popular adventure travel company that puts divers and photographers in the water with the world's largest and most dynamic predators.

Jillian Morris is a shark conservationist and award-winning photographer and videographer. She was born and raised in Maine, and after graduating with a degree in behavioral biology from the University of New England, she began traveling the world to work with sharks. Jillian worked as a research assistant and shark dive guide in the Florida Keys, the Bahamas, and Australia before moving to California to teach underwater photography and videography. Combining her science background with her passion for photography, Jillian joined the conservation media group OceanicAllstars. As executive director of that organization, Jillian has filmed for BBC, National Geographic, the Discovery Channel, Animal Planet, and many other networks. In 2012 she founded Sharks4Kids Inc., a nonprofit organization focused on creating the next generation of shark advocates through education, outreach, and adventure. Through this program Jillian is working to share shark education and conservation with students across the world. She is currently based on the island of South Bimini with her husband, Duncan Brake, and their adopted pit bull Lusca.

Duncan Brake is an Emmy-nominated cinematographer who has traveled, photographed, and filmed extensively throughout the world, from fighting bull elephant seals on the Antarctic Island of South Georgia to 4-meter tiger sharks in the crystal blue waters off Grand Bahama. After graduating in marine biology from Stirling University in Scotland, Duncan braved the elements of the South Atlantic for three years, working from the Falkland Islands down to Antarctica as a marine biologist and freelance cameraman. He then defrosted by venturing up to work as the Media Operations manager and assistant lab manager for the world-renowned Bimini Biological Field Station, where he gained thousands of hours of experience in the water working with, photographing, and filming sharks in their natural habitats. After spending over two and a half years at the Shark Lab, Duncan started working full time as a freelance camera operator. He has shot for BBC, CBBC, Discovery, National Geographic, Animal Planet, Discovery Canada, ITV, and many more. He is the founder of conservation media company OceanicAllstars and the media director for Sharks4Kids.